CARTRIDGE
MANUFACTURE

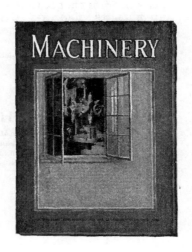

CARTRIDGE MANUFACTURE

A TREATISE COVERING THE MANUFACTURE OF
RIFLE CARTRIDGE CASES, BULLETS, POWDERS,
PRIMERS AND CARTRIDGE CLIPS, AND THE DESIGN-
ING AND MAKING OF THE TOOLS USED IN CONNEC-
TION WITH THE PRODUCTION OF CARTRIDGE CASES
AND BULLETS, TOGETHER WITH A DESCRIPTION
OF THE   PRINCIPAL OPERATIONS IN THE MANU-
FACTURE OF COMBINATION PAPER AND BRASS
SHOT SHELLS

By DOUGLAS T. HAMILTON

ASSOCIATE EDITOR OF MACHINERY
AUTHOR OF "ADVANCED GRINDING PRACTICE,"
"AUTOMATIC SCREW MACHINE PRACTICE,"
"SHRAPNEL SHELL MANUFACTURE"
"MACHINE FORGING," ETC.

FIRST EDITION

NEW YORK
THE INDUSTRIAL PRESS
1916

UF740
H3

PREFACE

The processes used in the manufacture of cartridges are of interest not only to the makers of munitions of war, but to every mechanic who is engaged in those industries in which the drawing and forming of metal in presses is of importance. The principles and operations pertaining to the production of metal cartridges are essentially the same as those used in hundreds of industries engaged in the manufacture of drawn objects for the most peaceful purposes; and while at the present time the interest in ammunition is paramount, the methods used for its manufacture differ in no essential from the methods employed in the ordinary metal working industries. The aim of this book is, therefore, not only to cover the various steps in the manufacture of one of the most important of munitions of war, but also to place on record approved methods in the drawing and forming of deep metal shells for any purpose. As the processes used in the manufacture of cartridges are probably more highly developed than those employed in most other fields of metal drawing, a book relating to these processes will no doubt be found of great value in many kindred fields.

In the field for which the book is especially intended, a record of the processes used in two leading plants for the manufacture of cartridges—one Government arsenal and one private concern—will require no special commendation. A great many firms are at the present time engaged in the manufacture of cartridges, and all mechanics occupied in this work will find a review of the methods of leading factories both of interest and value. It will enable them to compare their own methods with those used elsewhere, and to make improvements when the comparison is in favor of the methods described; and many will find suggestions for the introduction of entirely new methods of procedure. The author, therefore, has compiled this book with the hope that it will prove of service to the great number of men engaged in the manufacture of cartridges and in sheet-metal drawing work generally.

The book deals briefly with the design of various types of cartridges, the explosives used in them, and then more completely with the methods for drawing brass cartridge cases and making various types of bullets. A chapter is also included on the manufacture of primers and on the making of shot shells.

New York, January, 1916.   D. T. H.

CONTENTS

                                                                    PAGES
CHAPTER I.
     Development of Military Rifle Cart-
     ridges   1–9

CHAPTER II.
     Explosives Used in Rifle Cartridges and
     Primers ....'......   10–21

CHAPTER III.
     Manufacture of 0.22-Caliber Rifle Rim-
     fire Cartridges..   22–40

CHAPTER IV.
     Manufacture of Center-fire Cartridges..   41–60

CHAPTER V.
     Frankford Arsenal Method of Drawing
     Cartridge Cases..   61–74

CHAPTER VI.
     Making Dies for Cartridge Manufacture   75–87

CHAPTER VII.
     Making Spitzer Bullets.......... . .   88–104

CHAPTER VIII.
     Loading and "Clipping" Cartridges.... 105–116

CHAPTER IX.
     Manufacture of Shot Shells..... . ..   117–142

CHAPTER X.
     Manufacturing the French Military Rifle
     Cartridge   143–153

CHAPTER XI.
     Manufacture of Primers   154–162

CARTRIDGE MANUFACTURE

CHAPTER I

DEVELOPMENT OF MILITARY RIFLE CARTRIDGES

IN the early use of fire-arms for military purposes, the bullets, or balls, as they were then called, and the powder were carried separately. To load the musket, the powder was poured in from the muzzle, then wads were inserted and rammed, after which the spherical ball was inserted and held in place by ramming in more wads. This process was known as "muzzle loading," and the arm, as a "muzzle-loading musket." The first important advance was made early in the seventeenth century when Gustavus Adolphus, king of Sweden, gave instructions that gun powder be made up in the form of cartridges instead of being carried loose in flasks or bandoleers. The next advance was made early in the nineteenth century, when a percussion cap was devised. This made it possible to put up the correct charge of powder, ball, and ignition cap in the form of a paper cartridge, and subsequently made necessary the changing over of the musket from a muzzle to a breech loader.

The first use made of this cartridge was by a Prussian by the name of Von Dreyse, in his needle gun. The bullet used in this cartridge was egg-shaped, the base fitting into a shoe or wad of compressed paper, which carried the percussion cap in the opposite end. The wad was made of a larger diameter than the bullet, and fitted the rifling of the gun, thus imparting rotation to the bullet; but it fell to pieces upon emerging from the bore of the gun. The powder charge, bullet, and wad, were made up in a case of rolled paper, closed in at the rear and tied over the point of the bullet. The needle attached to the firing mechanism

1

pierced the base of the paper cartridge, and, by passing through the powder, ignited the cap. The method of explosion was, therefore, just the reverse of what it is at the present time.

Several other types of paper cartridges were afterwards made, but they did not prove successful. With paper cartridges, it was practically impossible to seal the breech of the gun to prevent the escape of gases without resorting to a specially constructed breech mechanism, and endeavors were made to produce a cartridge which would fill these requirements.

The first brass cartridge case was invented by Col. Boxer, in 1865, and was used in the Enfield muzzle-loading rifle which was converted into a breech loader, adopting the principle patented by an American, Jacob Snider. This cartridge was made from built-up sections of rolled brass, and carried a percussion cap in its head. Cartridge cases made from drawn brass were not introduced into England, except for machine gun use, until after the Egyptian campaign of 1885.

Cartridge Cases. — Cartridge cases made from drawn brass are produced by first cutting out a disk of the required diameter and thickness, forming it into a cup, and then, by successive redrawing operations, reducing its diameter and increasing the length. Upon the advent of the small-bore, high-power rifle, the shape of the cartridge case was changed from the straight-body type to the bottle shape. This was done in order to increase the size of the powder chamber without increasing the length of the case. Cartridge cases were first made with flanges, but, upon the application of clips and chargers, the rim was ·omitted and a groove cut in the head instead. This allows the cartridge cases to be packed closer together, and also prevents jamming in the magazine action.

The cutting of a groove in the rim of the head, as well as the substitution of smokeless powder in the place of black powder, made necessary a change in the shape of the head of the case. The first solid drawn cartridge cases were made as shown at *A*, Fig. 1. The case was drawn up from a comparatively thin blank, and to form the pocket for the primer the metal was simply forced into the case as shown. A case of this shape could not satisfactorily be grooved, as the head would thereby be considerably weakened. An improvement was made, therefore, consisting in making the case from a much thicker blank, and, instead of bending in the metal at the head to form the primer pocket, the metal was simply displaced so that the head remained solid, as shown at B, Fig. 1. This type of cartridge case has now been adopted by most of the principal governments.

Bullets. — As has been previously mentioned, the first bullets were spherical in shape, as shown at A, Fig. 2, and were fired from smooth-bored muskets. Upon the invention of the rifled gun early in the seventeenth century, the first successful bullet used was devised by a Swiss, Major Rubin, and is shown at B, Fig. 2. This was still spherical in shape, but had a copper belt a cast completely around, which "took" the rifling grooves in

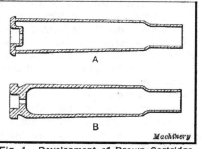

Fig. 1. Development of Drawn Cartridge Cases

the gun. This bullet, however, was very erratic in flight, and the next step was to make it egg-shaped, as shown at C. To provide for the expansion of the bullet, the breech was equipped with a taper steel pin against which the bullet was rammed from the muzzle. The ram-rod was provided with a cavity in its end of the same shape as the bullet, and, by driving the latter against the tapered plug, the base end was expanded into the rifling grooves.

The next improvement, made in 1836, which, however, was not generally adopted, was the Grenner bullet shown at D. This was made with an oval head and a flat base, made hollow, and a tapered plug of wood inserted in the cavity. Upon the explosion of the powder in the chamber of the gun, the plug was driven into the soft lead bullet, expanding it to fill the rifling grooves. Captain Minie, of the French army, in 1849, produced a bullet of the shape shown at *E*, in which the wooden plug was replaced by a hemispherical iron cup. This cup, as was the case with the wooden plug, served to expand the bullet into the rifling grooves.

The first lead bullets of the shape shown at *F* were made from pure lead, and, to avoid fouling of the bore of the gun, were covered with waxed paper *b*. The Snider bullet of this type was 0.489 inch in diameter, and had a length of 1.8

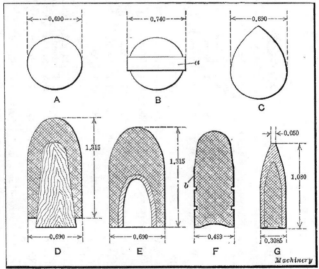

Fig. 2. Early Development of Bullets for Military Cartridges

calibers, whereas, the Martini-Henry was 0.450 inch in diameter, and had a length of 2.8 calibers; both weighed 480 grains. Subsequent developments consisted in a gradual reduction in diameter and increase in length up to and exceeding four calibers. The bullets were also hardened by adding antimony and tin. Bullets made from lead and antimony, however, were still found to be too soft to be fired from high-powered rifles, so that the next development consisted in covering the lead core with a metal jacket, as at *G*.

PRINCIPAL DIMENSIONS, WEIGHTS, ETC., OF MILITARY RIFLE CARTRIDGES

Country	CARTRIDGE								BULLET				
	Length of Case, Inches	Total Length, Inches	Weight, Grams	Shape of Head	Propellant	Wt. of Charge, Grains	Muzzle Vel. Ft. per Sec.	Chamb. Pres., Tons Sq. In.	Shape of Nose	Material of Envelope	Length, Inches	Diam., Inches (Max.)	Weight, Grains
Austria-Bulgaria	2.06	3.0	445.0	Rim	Nitrocellulose	42.44	2034	19.7	Round	Steel Lubricated	1.24	0.3298	244.0
Belgium	2.10	3.055	441.0	Rimless	Nitrocellulose	37.0	2034	19.7	Round	Cupro-nickel	1.205	0.31	219.0
Canada	2.21	3.05	415.0	Rim	Cordite	31.5	2060	15.5	Round	Cupro-nickel	1.125	0.311	215.0
Denmark	2.28	3.0	460.0	Rim	Nitrocellulose	33.95	1965	15.1	Round	Cupro-nickel	1.187	0.323	237.0
France	1.917	2.95	415.0	Rim	Nitrocellulose	46.2	2380	17.75	Pointed	No Envelope, Solid Copper-zinc	1.625	0.327	198.0
Germany	3.239	3.189	369.9	Rimless	Nitrocellulose	48.4	2882	17.75	Pointed	Steel, Coated with Nickel	1.105	0.323	154.5
Gt. Britain	2.21	3.05	415.0	Rim	Cordite	31.5	2060	15.5	Round	Cupro-nickel	1.125	0.311	215.0
Greece	2.11	3.05	348.0	Rimless	Nitrocellulose	56.0	2223	20.18	Round	Steel, Coated with Cupro-nickel	1.124	0.263	159.3
Holland	2.11	3.05	388.0	Rim	Nitrocellulose	36.26	2433	17.75	Round	Steel, Coated with Cupro-nickel	1.23	0.2687	162.0
Italy	2.06	3.0	331.8	Rimless	Ballistite	30.09	2395	17.1	Round	Cupro-nickel	1.182	0.266	163.0
Japan	2.0	2.98	348.5	Semi-Rimless	Nitrocellulose	32.0	2390	17.1	Round	Copper	1.28	0.26	162.9
Portugal	2.35	3.26	.....	Rim	Nitrocellulose	31.8	2347	17.75	Ogival	Steel, Coated with Cupro-nickel	1.28	0.320	155.3
Roumania	3.10	3.05	350.0	Rim	Nitrocellulose	36.0	2400	17.75	Round	Cupro-nickel	1.244	0.2687	162.0
Russia	2.11	3.025	363.0	Rim	Pyroxiline	37.0	1985	17.47	Round	Cupro-nickel	1.194	0.308	214.0
Spain	2.22	3.08	373.5	Rimless	Nitrocellulose	38.35	2296	22.3	Round	Cupro-nickel	1.21	0.2843	172.8
Switzerland	2.11	3.043	434.0	Rimless	Nitrocellulose	50.7	1920	17.1	Round	Plated Steel over Point, only Steel, Coated	1.18	0.319	212.5
Turkey	2.21	3.07	416.0	Rimless	Nitrocellulose	40.2	2066	19.7	Round	Steel, Coated with Cupro-nickel	1.212	0.311	211.3
U. S. A.	1.95	3.329	392.0	Rimless	Pyrocellulose	50.0	2600	19.78	Pointed	Cupro-nickel	1.08	0.3085	150.0

Modern Military Rifle Cartridges. — The military rifle
cartridges used by some of the principal governments are
shown in Fig. 3. This particular grouping has been
selected because of the variation in design. The main difference is in the shape of the bullet, and the total length.
The bullets, it will be noted, are all of the coated or jacketed
types, with the exception of the French; this is solid
and is made from a copper-zinc alloy. The bullet used in
the United States cartridge is a cupro-nickel jacket over a
coating of lead, hardened with antimony. The same re-
marks apply to the British, Italian, and Russian bullets.
The German bullet has a lead filling and is coated with steel,
nickel-plated; the Austrian bullet has a steel jacket which is
lubricated. Further details regarding the cartridges
adopted by the principal governments are given in the ac-
companying table.

Cupro-nickel jacketed bullets are generally employed for
military rifles and are used by the Belgian, British, Cana-
dian, Danish, Italian, Roumanian, Russian, Spanish, and
American governments. The German, Greek, Dutch, and
Turkish governments use steel envelopes coated with cupro-
nickel; Austria uses greased steel, and Japan, copper. Bullets coated with cupro-nickel are likely to set up metallic
fouling in the bore of the gun, consisting of streaks of
metal which adhere to the lands and grooves in the bore.
Bullets with greased steel envelopes do not appear to cause
metallic fouling, but they wear away the rifling in the gun
much quicker.

The bullet used in the Swiss rifle cartridge is of a peculiar construction. The body is made of a hard lead alloy,
provided with a nickel-plated steel envelope covering the
point only, the remainder of the bullet being covered with
paper lubricated with vaseline. The lower portion of the
bullet which enters the cartridge case is smaller in diameter
than the jacketed portion. The wounding power of this
bullet is great, but its velocity is not as great as those provided with the full envelope.

Since the outbreak of the present war, considerable
changes have been made in the bullets used by the various

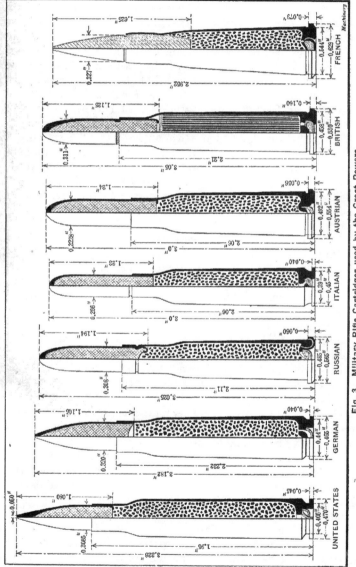

Fig. 3. Military Rifle Cartridges used by the Great Powers

participants. Up to the present war, Great Britain used a
round-pointed bullet, but has now adopted the pointed type.
Pointed bullets are being gradually adopted by all of the
governments, because the sharp head offers less resistance
to the air than does the rounded head. For this reason, the
pointed bullet always has a higher velocity, all other factors
remaining the same. To further reduce air resistance, the
French bullet is made with a slightly tapered rear portion.
This reduces the vacuum formed at the base of the bullet,
and, hence, makes possible an increase in velocity.

Cartridge Clips and Chargers. — Soon after the adop-
tion of the rifled gun, steps were taken to increase the
rapidity of fire. Weapons known as repeaters were the first

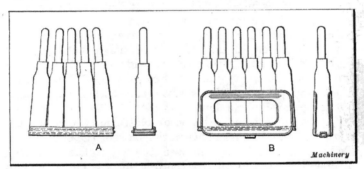

Fig. 4. Cartridge Chargers and Clips

really successful rifles used. These contained a supply of
cartridges which were held in a tube located beneath and
parallel with the barrel. The cartridges lay nose to base
in this tube, and, upon the operation of the breech mechanism, the exploded cartridge was ejected and a fresh one
inserted automatically. The earliest weapons of this type
were the Spencer and Henry rifles patented about 1860.

With this type of rifle, it was necessary, of course, to fill
up the magazine by placing one cartridge in it at a time, and,
to facilitate loading, devices called quick loaders were developed. The first successful quick loader was known as
the "Kruka" and was developed in Russia during the Russo-
Turkish war, in 1878. This device, which was the forerun-

ner of the present charger, was attached to the side of the stock and held ten cartridges in pockets with the base up.

The next step was made by Austria, in 1886, in which year the Mannlicher rifle was adopted by that government, and was provided with a box magazine. This box was located in the rear of, and below the entrance to, the gun chamber, and held the cartridges in a horizontal plane. A spring-actuated platform pushed the cartridges upwards when the bolt was operated. Another improvement made was the application of a spring clip to facilitate the loading of the cartridges into the magazine. This was made of sheet steel and held five cartridges. It was pressed into the magazine where it was retained by a catch until the cartridges were used, after which it fell out through an opening in the bottom of the magazine.

Chargers are used simply for convenience, in carrying and loading the cartridges in the magazine; they are placed above the magazine, the cartridges swept out by the thumb, and the empty charger thrown away. Clips, on the other hand, are placed in the magazine with the cartridges and are generally held down by a catch. The cartridges are fed up by the magazine lever, or platform, which is made narrow enough to pass between the sides of the clip. When all the cartridges have been used, the clip falls out through an opening provided in the bottom of the magazine. A common form of charger is shown at A, Fig. 4. This is known as the Mauser charger and consists of a thin undulating flat spring that presses the cartridge forward so that the extractor grooves in the head of the case bear firmly against the ribs on each side of the charger. Chargers are generally made to hold five cartridges, but some hold six. A greater number of cartridges than this are likely to jam when being swept out of the charger into the magazine.

The clip shown at B, Fig. 4, differs from the charger in that it is placed directly in the magazine and remains there until all of the cartridges are extracted, whereupon it drops out through an opening in the bottom of the magazine. Clips are also made with extended sides to support the cartridges and prevent them from clogging in the magazine.

CHAPTER II

EXPLOSIVES USED IN RIFLE CARTRIDGES
AND PRIMERS

EXPLOSIVES, in general, may be defined as substances, either solid or liquid, which, upon the application of heat or other causes that set up a chemical action in them, are capable of being converted instantaneously into gases, occupying a much greater volume than the original substance. Explosives may be divided into three general classes:

1. Low, also known as progressive or propelling explosives.

2. High, or detonating explosives.

3. Fulminates, or detonators.

The first class includes black gun powder, smokeless powder, and black blasting powders. The second includes dynamite, nitroglycerine, guncotton, picric acid, etc., and the third includes mercury fulminate and chlorates.

Black Gun Powder. — The early history of black gun powder is very indefinite. As far as can be ascertained, it was first produced in England in the thirteenth century, but was not used to any great extent until the latter part of the eighteenth century. Berthold Schwarz, a German monk, and Roger Bacon, an English friar, are both credited as being the original inventors of gun powder. Roger Bacon, in a book published by him in 1242, makes reference to an explosive mixture containing saltpeter that "makes a noise like thunder and flashes like lightning." The three principal ingredients of black gun powder are saltpeter, charcoal, and sulphur in about the following proportions: saltpeter, 75 parts; charcoal, 15 parts; sulphur, 10 parts.

The relative proportions of saltpeter, charcoal, and sulphur used vary in different countries, and have also been changed from time to time since the first gun powder was manufactured. The advance seems to be along the line of increasing the proportion of saltpeter and relatively decreasing the proportions of charcoal and sulphur.

Charcoal is the chief combustible in powder. It must burn freely, leaving as little light ash or residue as possible. It must be friable and grind into a non-gritty powder. The materials from which powder charcoal is made are dogwood, willow, and alder. Dogwood is mainly used in the manufacture of powder for rifle cartridges, etc. Powder made from dogwood charcoal burns more rapidly than that made from willow, etc. The wood, after cutting, is stripped of bark and allowed to season for two or three years. It is then segregated into lots of a uniform size and charred in cylindrical iron cases or slips, which are placed into slightly larger cylinders set in the furnace. The slips are provided with openings for the escape of the gases. The rate of heating, as well as the absolute temperature attained, has a more or less marked effect on the product. A slow rate of heating yields more charcoal and a high temperature reduces the oxygen and hydrogen in the final product. When heated for seven hours to about from 800 to 900 degrees C. (1472 to 1652 degrees F.), the remaining hydrogen and oxygen amount to about 2 per cent and 12 per cent, respectively. The time of charring, as a rule, is from five to seven hours. The slips are then removed from the furnace and placed in a larger iron vessel where they are kept comparatively air-tight until quite cool. The charcoal is then sorted and stored for some time before grinding. The grinding which follows consumes several hours, and as the "dust" passes out from the rollers it is sifted on a rotating wheel or cylinder of fine-mesh copper wire gauze. The sifted charcoal powder is again stored for some time before using and is kept in closed iron vessels.

The sulphur used in the manufacture of gun powder is generally obtained from Sicily. For complete purification, the sulphur is first distilled and then melted and cast into molds. It is afterwards ground into a fine powder and sifted, as in the case of the charcoal.

Saltpeter—the common name for potassium nitrate—is the oxygen provider in gun powder. This is a valuable salt and, in many localities, is found in caverns or caves in calcareous formation, but the chief commercial source of

this salt is the soil of the tropical regions, especially of the
districts in Arabia, Persia, and India, where the nitrate is
found as an efflorescence upon the surface of the ground, or
in the upper portions of the soil itself. The niter is extracted by treating the earth with water and obtaining it in
an impure state by evaporating solutions. The crude product is purified by successive re-crystallizations. Niter is
soluble in water and much more so in hot than cold water.
The crude material is dissolved almost to the saturation
point in boiling water and, on filtering and cooling this
liquor to 30 degrees C. (86 degrees F.), almost pure niter
crystallizes out, most of the usual impurities still remaining
in solution. By rapidly cooling and agitating the niter solution, crystals are obtained of sufficient fineness for the
manufacture of powder without special grinding. Niter
contains nearly 40 per cent of oxygen, by weight, five-sixths
of which is valuable for combustion purposes.

Manufacture of Black Gun Powder. — The materials
used are weighed out separately, mixed by passing through
a sieve, and then uniformly moistened with a small quantity
of distilled water while on the bed of the incorporating mill.
This consists of two heavy iron wheels mounted so as to
run in a circular bed. The incorporation of the material
requires about four hours. The mechanical action of the
rollers on the powder paste is a double one, not only crush-
ing, but mixing, by pushing forward and twisting sideways.
The pasty mass is deflected so that it repeatedly comes
first under one roller and then under the next by scrapers
which are set at an angle to the bed and follow each wheel.

Although the charge is wet, while in this condition, it is
possible for it to be fired either by the heat developed by
roller friction, by sparks and foreign matter, or by bits of
stone, etc. The mills are, therefore, provided with drench-
ing apparatus so that in case of one mill firing, this mill
and the one next to it will be drenched with water from the
cistern or tank immediately above the mill. The product
from the incorporation mill is termed a "mill cake."

Upon the completion of this operation, the ingredients
are in a damp state, and are then dried and shaken to

separate them. After the pressing of the incorporated powder into a "press cake," it is broken up or granulated bv
suitable machines. These are generally provided with Tobin-
bronze rollers which revolve, and through which the cake
passes. The "meal" is then pressed in a box between Tobin-
bronze plates, in layers, by means of hydraulic pressure,
the result being a "press cake." The next operation consists of granulating or converting the press cake into grain
powder. The granulating machine consists of a strong
Tobin-bronze framework carrying two pairs of toothed and
two pairs of plain Tobin-bronze rollers, and furnished with
the necessary sifting screens. The grains are next separated from the dust by passing the material through a sloped
revolving reel covered with 20-mesh copper wire. The
grains are then polished by rotating in drums alone or with
graphite, which adheres to and coats the surface of the
grain. This process is generally followed with powders
intended for small arm or moderate small ordnance use,
and graphite is seldom used, the required glaze being obtained merely by the friction of the grains.

Gun powder is generally made up into different shapes of
various sizes. Prisms or prismatic powder is made by
breaking up the press cake into a moderately fine state while
still moist and pressing a certain quantity into a mold. The
mold generally employed consists of a thick plate of bronze
having a number of hexagonal perforations. Accurately
fitting plungers are so applied to the holes in the plate
that one enters the top and the other the bottom. After
the desired pressure has been applied, the top plunger is
- withdrawn and the lower one pushed upward to eject the
"prism" of powder. In 1860, General T. J. Rodman sug-
gested perforating the powder prisms in order to increase
the rate of burning. The axial perforations in prism powder are made by small bronze rods which pass through
the lower plunger and fit corresponding holes in the upper
one. Other shapes to which gun powder is made are accomplished in a somewhat similar manner. The "prismatic"
powder or other shapes, however, are not used in small
arms. Here the ordinary grain powder has been almost

universally adopted. Black gun powder, of late years, however, has been substituted by smokeless powder which gives much better results.

Smokeless Powder. — Smokeless powder, as its name implies, produces no smoke following explosion, but this result is seldom realized due to the action of certain ingredients used as necessary adjuncts. The invention of smokeless powder is attributed to a German chemist, Schoenbein, who, in the year 1846, discovered a substitute called "cotton powder" or guncotton, which he proposed to take the place of gun powder as it had certain advantages over the latter, due to burning without any noticeable residue and, consequently, with little smoke. Smokeless powder, however, was not satisfactorily produced in large quantities until some time later, and several very serious accidents occurred before its manufacture was brought to a commercial basis. General Von Lenk, of the Austrian Artillery, succeeded in improving considerably upon the method of manufacture and purification. He so altered the mechanical condition of the guncotton as to modify its rate of combustion in the air and, therefore, to render its application possible for military purposes. It was finally adopted by the Austrian government in 1862.

Von Lenk's system consisted in making guncotton from yarn and thread of various sizes and degrees of compactness, spun from long staple cotton, and then twisting and plating the guncotton yarn in different ways. It was then wound more or less tightly over cones or spindles of wood or paper, and also woven into cartridges of various shapes and sizes.

A still further improvement was made, in 1863, by Sir Frederick Abel, whereby a practically complete purification was obtained and the material was converted by means of reduction to a fine state of division and subsequent compression. In this form, guncotton was introduced for torpedo and mine charges, but, to develop a full explosive effect, it was necessary for the charge to be strongly confined, which greatly reduced its range of application. All necessity for confinement ceased, however, when it was discovered, in 1869, that compressed guncotton, either dry or wet, could be fully detonated by fulminate of mercury in a totally unconfined state. The first application of guncotton for sporting rifle cartridge use was patented by Schultze, in 1867. This powder consisted originally of nitrated wood fiber impregnated with nitrates and chlorates. It was not entirely smokeless and was never used for military purposes.

In 1885, upon the introduction of small caliber magazine rifles, a great improvement was made in the methods of treating nitrated cellulose, so that these substances could be handled with safety. The improvement consisted in converting it into a substance absolutely devoid of all porosity, and was obtained by subjecting the nitrated cellulose to the action of suitable solvents which would gelatinize it and, upon evaporation, leave a compact homogeneous non-porous material capable only of burning from the exterior towards the center. The gelatinized material, before evaporation, could, while in a plastic state, be rolled or pressed into sheets or cords, or any other desired form. The first smokeless powder for military rifle use was invented by a French chemist, Vieille, in 1884, and was used in the Lebel rifle. It was originally a mixture of nitrocellulose and picric acid. The picric acid, however, was subsequently abandoned and the powder now consists of a gelatinized mixture of soluble and insoluble nitrocelluloses.

A still further improvement was made in the manufacture of smokeless powder by a Swedish engineer, Alfred Nobel, in 1886. This powder was patented, in 1888, under the name of "ballistite" and is still employed in some countries. This is a product combining nitrocellulose and nitroglycerine, which are mixed by rolling between hot rollers, the union being promoted by employing camphor. The camphor, however, did not remain a constant ingredient, owing to its volatility, and resulted in alterations in the ballistic properties of the material. This is a grave defect in an explosive, especially when it has to stand exposure in various climates. To overcome this objection, camphor was abandoned and guncotton, instead of soluble nitrocotton, was used to obtain an explosive of uniform composition.

The final result was the production of a smokeless propulsive explosive consisting essentially of nitroglycerine and guncotton, incorporated and gelatinized by the aid of a solvent. A small proportion of mineral jelly was incorporated in the preparation to prevent "metallic fouling" in the magazine rifle. This was also found to improve the stability of the powder under various climatic conditions. This explosive is known by the name of "cordite," owing to the cordlike form it assumes in manufacture. The first mixture used in this explosive was composed as follows: Nitroglycerine, 58 per cent; guncotton, 37 per cent; mineral jelly, 5 per cent.

After a considerable series of experiments were made, it was found that this explosive had serious erosive qualities, particularly in heavy ordnance, and it was found necessary to reduce the percentage of nitroglycerine. The following composition was substituted: Nitroglycerine, 30 per cent; guncotton, 65 per cent; mineral jelly, 5 per cent. Within the last few years, however, cordite has been substituted, to some extent, by nitrocellulose and pyrocellulose smokeless powders. These do not have such a deteriorating effect upon the bore of the gun and, in many other respects, are better propellants.

Manufacture of Smokeless Powder. — The manufacture of practically all smokeless powders is conducted generally along the same lines. The principal operations consist of incorporation and gelatinization of the ingredients by means of a solvent, or by heat and pressure; converting the gelatinized material into the desired form by pressure, etc.; and the elimination of the solvent. The body of all smokeless powders is cotton, and the chief ingredients, nitric and sulphuric acids. For the first operation, or the incorporation and gelatinization of the ingredients, a kneading machine is usually employed. This consists of an iron box located on suitable supports, open at the top and provided with removable covers, the bottom being shaped to form two semi-circular troughs in each of which a spindle carrying propeller-shaped blades revolves. The spindles turn in opposite directions, one moving at twice the speed of the

other. The blades revolve in close proximity to the bottom
of the machine, and the material is continually being
squeezed between the blades and the bottom of the machine,
and between the blades themselves as they approach each
other along the center line. The action is, in fact, a kneading one, and the machine is very similar to those used for
making dough and many other similar processes.

In the manufacture of nitrocellulose, the cotton used is
generally the short fiber which is detached from the cotton
seed rather late in the process of removal. After being
treated and purified, the cotton is run through a picker
which opens up the fiber and breaks up any lumps; it is
then thoroughly dried and ready for nitration. The most
commonly used method of nitration is to put the cotton
into a large vessel nearly filled with a mixture of nitric
acid and sulphuric acid. Sulphuric acid is used to absorb
the water developed in the process of nitration which would
otherwise too greatly dilute the nitric acid. After a few
minutes immersion, the pot is rapidly rotated by machinery
and the acid permitted to escape. The nitrated cotton is
then washed in a preliminary way and removed from the
nitrater, and repeatedly washed and boiled to remove all
traces of free acid. In the process of nitration, the cotton
does not change materially in its appearance, but becomes
somewhat rough. As the keeping qualities are dependent
upon the thoroughness with which it is purified, the specifi-
cations for powder for the U..S. army and navy require
that the nitrocellulose shall be given, at this state of manufacture, at least five boilings with a change of water after
each boiling, the total time of boiling being forty hours.

Following this preliminary purification of the nitrocellulose, it is cut up into shorter lengths by being repeatedly
run between cylinders carrying revolving knives, as previously described. This operation is necessary as cotton
fibers are hollow, thus making it very difficult to remove
traces of acid from the interior unless cut up into very
short lengths. After being pulped, the nitrocellulose is
given six more boilings and a change of water after each,
followed by ten cold water washings. The completed mate-

rial is known as guncotton or pyrocellulose. Before adding the solvent, the pyrocellulose must be completely freed from water. This is partly accomplished in a centrifugal wringer, but is completed by compressing the pyrocellulose into a solid block, then forcing alcohol through the compressed mass. Some of the water is thus forced out ahead of the alcohol and the remainder is absorbed by the alcohol, the operation of forcing it through the block being continued until pure alcohol appears. Ether is added to the pyrocellulose thus impregnated with alcohol, the relative proportions being about two parts of ether to one part of alcohol, by volume. The amount of mixed solvent added varies between 85 and 110 per cent of the weight of the dried pyrocellulose. After the ether has been thoroughly incorporated in the kneading, the material is placed in a hydraulic press, which forms it into cylindrical blocks about 10 inches in diameter and 15 inches long. In this operation, the pyrocellulose loses the appearance of cotton and takes on a dense horny appearance, forming what is known as a "colloid." The colloid is transferred to a finishing press where it is again forced through dies and comes out in the form of long strips or rods which are cut to grains of the length required. The grains are then subjected to a drying process which removes nearly all of the solvent and leaves the substance in a suitable condition for use. The drying process is a lengthy one, requiring from four to five months for the larger grain powder. This process of drying, etc., has, of late, been improved upon by the DuPont Powder Co., and the time required to manufacture nitrocellulose has been very greatly reduced. Upon completion, the powder is blended and packed in air-tight boxes.

Classification of Smokeless Powders. — The form of grain in which smokeless powder is made differs largely in various countries. Some use flake formation, whereas others use long strings. The flake form of smokeless powder is used largely in sporting cartridges, but in military cartridges extensive use is made of the cord type, or in other words, "cordite." Some countries use smokeless powder in the form of tubes or cylinders, either solid or tubular.

Sporting powders are required to burn more quickly and are generally granular in form, like old gun powder, or in the form of very thin flakes. The color varies considerably and depends, to a great extent, on the amount of non-explosive material added in the ingredients. Pure nitrocellulose powders are, as a rule, gray or yellow; powders in which nitroglycerine is present vary in color from light yellow to a deep brown; sporting powders sometimes contain coloring matter and are frequently coated with graphite, which gives them a silvery gray appearance. The surface of the flake, tube, or cord powder is usually smooth and hard. The texture is horn-like if the powder is made from nitrocellulose, but softer and of the consistency of India rubber if containing nitroglycerine. The density of powders varies according to the ingredients and methods of manufacture. Unless the powders contain ingredients soluble in water, such as metallic nitrates, they are unaffected by dampness, and do not absorb any appreciable amount of moisture. To insure uniformity in the ballistic properties of the finished explosives, the blending of different batches is resorted to. The smokeless powders in use for military small arm cartridges are conveniently classed under two main headings; 1. Powder composed of nitrocellulose, either alone or with small quantities of other explosive or non-explosive ingredients. 2. Powder containing nitroglycerine in various proportions as a further ingredient. The military small arm powders of Class I are those used by Argentine, Austria, Belgium, Brazil, Bulgaria, China, Colombia, Denmark, France, Germany, Holland, Japan, Mexico, Portugal, Roumania, Russia, Servia, Spain, Switzerland, Turkey, United States, and Uraguay. Class II applies to powders used by Great Britain, Italy, and Norway.

Mercury Fulminate. — The detonating substance used in primers and detonating caps is comprised chiefly of mercury and fulminic acid. The salts of this acid are very explosive and, when mixed with mercury, form a very suitable substance for use in primers. The first fulminate prepared was the fulminating silver discovered by L. G. Brugnatelli, in 1798. He found that if silver were dissolved in

nitric acid and the solution added to spirits of wine, a white,
highly explosive powder would be obtained. This substance
is distinguished from the black fulminating silver discovered by C. L. de Berthollet, in 1788, which was obtained
by acting with ammonia on precipitated silver oxide. The
next salt to be obtained was mercuric salt, which was prepared, in 1799, by Edward Charles Howard, who substituted mercury for silver in the original Brugnatelli's process. A similar method was that of J. Von Liebig who, in
1823, heated a mixture of alcohol, nitric acid, and mercuric
nitrate. The use of mercuric fulminate as a detonator dates
from about 1814, when the explosive cap was invented.
This is still the most common detonator, although it is
usually mixed with other substances. For a percussion cap
the British service cartridge contains 6 parts of fulminate,
6 parts of potassium chlorate, and 4 parts of antimony sul-
phide. The most common fulminate used at the present
time is made by dissolving the metal mercury in strong
nitric acid and pouring the solution into alcohol. After an
apparently violent reaction, a mass of fine grain crystals
of fulminate of mercury is produced. The crystalline powder that is produced is washed with water to free it from
acid and, because of its extreme sensitiveness, is usually
kept soaked with water or alcohol until needed. One must
be very careful in handling this material; the mixing is
done on rubber-covered tables, the operator wearing rubber
shoes and gloves. One peculiar feature about this mixture
is its attack upon the teeth, causing them to blacken and
decay rapidly. For use in primers, it is usually mixed
with ground glass, the latter forming the friction element
in this explosive. It can be detonated by percussion, friction, or heat. On account of its great specific gravity, a
small volume of this substance develops a large volume of
gas. According to the usually assumed reaction, the gas
developed occupies, at ordinary temperature, a volume more
than 1340 times that of the original material. Because of
the large amount of heat developed in reaction, the volume
of the temperature at reaction is very much greater. It is
estimated that a pressure as great as 48,000 atmospheres

(700,000 pounds per square inch) is produced by the detonation of mercury fulminate.

Chlorate of Potassium. — Fulminate of mercury for use in cartridge primers has been largely superseded by chlorate of potassium, because of the tendency of the former to cause fouling in the bore of the gun. Chlorate of potassium as a detonating composition is used in primers for cartridges of the British, American, and some of the foreign types. The common mixture of chlorate of potassium includes the following ingredients:

Ingredients	Per cent
Chlorate of Potassium..	49.6
Sulphide of Antimony...........	25.1
Glass (ground)	16.6
Sulphur	8.7

This composition is much less dangerous to handle than fulminate of mercury, both during the manufacturing processes and also when loaded into the primer.

CHAPTER III

MANUFACTURE OF 0.22-CALIBER RIFLE RIM-FIRE CARTRIDGES

IN the manufacture of rim-fire cartridges, of which the 0.22 long-regular type of cartridge may serve as a suitable example, the important operations are as follows: Blanking, cupping, drawing, annealing, washing, drying, trimming, heading, priming, loading, and inspecting. This chapter describes the practice of the Dominion Cartridge Co. in making cartridges of this size and type. A description of the manufacture of the lead bullets used in this size of cartridge is also included.

Cupping and Annealing. — The first operation in the manufacture of rim-fire (RF) cartridges is to produce a cup of the required size by means of a combination blanking and cupping punch and die. The cartridges are usually blanked and cupped at a rate of from four to six for each stroke of the press. The cup is made from a blank sheet of copper of the required width and thickness. Before any drawing operation can be performed on the cups, they must be annealed to make them ductile. The cups are placed in a cylindrical cast-iron drum, shown in Fig. 1, which has holes, smaller in diameter than the smallest cups annealed in it, drilled around its periphery. These holes permit the heat to permeate, thus annealing the cups rapidly. The cylindrical drum is provided with a slide or door, which is forced in when the cups are inserted. The drum is then rolled into the furnace, where it is rotated by means of a chain and sprocket, driven by an overhead shaft. The door A, which is shown in the upper position in Fig. 1, is then brought down. Before the cups are inserted, the drum must be heated to a cherry red. Then the cups are put in and allowed to remain for a period of from thirty to forty minutes. After the cups have remained in the drum the specified time, the front door is raised and the drum rolled out. A truck with a pan located on it is then rolled in front of the annealer, the slide taken out, and the cups dumped into this pan, which is filled with lukewarm water.

Washing the Cups.—In the annealing process, a scale is formed on the cups, which would be detrimental in drawing; therefore, before any drawing operations can be carried out, they must be washed. To accomplish this, the cups are dumped into revolving tubs, as shown in Fig. 2. These tubs are driven by a shaft located beneath them, through bevel gears. A clutch is also provided, so that any one of the tubs may be stopped, if desired, independently of the others. By rotating these tubs, the cups are made to rub against one another, thus helping to remove the

Fig. 1. Cylindrical Drum In which the Cups are Annealed

scale. The rotation of the cups and the pouring of water on them is not sufficient to remove the scale, so they are immersed in a solution composed of sulphuric acid and water, mixed together in the following proportions: Water, 48 pints; sulphuric acid, 1/4 pint. This solution is used for the first washing and removes the scale. When the cups look quite clean, a plug is removed and the acid solution washed off. Then the plug is inserted and another solution composed of pearlash soda, soft soap, and water is mixed together in the following proportions: Water, 48 pints; soft soap, 1 pint; pearlash soda, ¼ pint. The cups are rotated in this solution for about twenty minutes, after which they are rinsed with warm water. A sieve is located over the hole for the plug so that the cups cannot fall out. When the cups have been thoroughly rinsed and the water drained off, they are put in sieves, which, in turn, are placed in a cupboard, where they are held in racks. Steam pipes are located beneath these racks, so that the cups are quickly dried.

Fig. 2. Tubs used in Wash-   Fig. 3. Vertical Header with
ing the Cups after Annealing   Automatic Feed

Drawing Operations. — When dry, the cups are transferred to the drawing department. Here they are put in a pan, from which the operator removes them by means of a vulcanite plate. This plate has a series of holes drilled through it, which are bell-mouthed on the top, and are slightly larger in diameter than the cups. A thin sheet-steel plate, bent up on two sides to fit the vulcanite plate, is slipped under it when shaking in the cups. When the cups are shaken into the plate, the operator places it on the table

A of the double-headed friction-dial drawing press shown in Fig. 4, after which the sheet-steel plate is removed, and the vulcanite plate lifted up, leaving the cups standing on the table *A*. The operator then shoves the cups from the table onto the friction dial *B*, which carries them to the dies. These friction dials are driven through bevel gears by a round belt, which is connected through three grooved pulleys to the main driving pulley. The cups pass between the guard *C* and spring *D*, the latter being vibrated by the action of the revolving dial, which keeps the cups in constant motion, thus arranging them in single file. The guard *C* and the spring *D* approach each other as they near the

Fig. 4. Double-headed Friction-dial Drawing Press used for the Various Drawing Operations

dies to within a distance equal to the diameter of the cups. The cups are carried from the dial to the dies by a finger *E* connected to the bellcrank *F*, which, in turn, is operated by the cam *G* on the end of a vertical shaft, driven from the crankshaft through bevel gears.

The action of drawing the cups through the dies is clearly shown in Fig. 5. The cup is carried out by the finger and placed over the die *i*. The punch *h* then descends in it, as shown at *A*, forcing it down into the die, as shown at *B*. In the latter position, the case is given a mushroom shape, and is then forced through the die, as at *C*, reducing it in

diameter, increasing its length and decreasing the thickness
of its walls. As the cup is forced through the die, it expands slightly away from the punch, and on the upward
stroke of the press it catches on the bottom of the die, and
is stripped from the punch, dropping into a box placed
under the machine. Only one die is shown in the illustra-

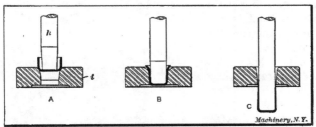

Fig. 5. Diagram showing the Action of Drawing the Cups

tion, but in actual practice two drawing dies are used, one
on top of the other. The lower die is not bell-mouthed, but
is slightly tapered. In the drawing operations it is necessary to lubricate the cups, so that they will not stick to the
die or punch, and for this purpose a lubricant composed of
soft soap and water is used. This is placed on the cases
while they are located on the dial, by the operator, who

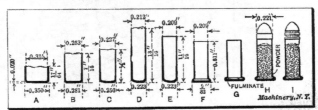

Fig. 6. Illustration showing the Transition from the Cup to
the Finished Cartridge

spreads the lubricant over them by means of a small cup
fastened to a handle. The punch for the drawing operations is made slightly tapered, being smaller at the lower
end, as shown in Fig. 5. This is necessary, as the explosive
material is placed near the head of the case, thus requiring
the walls to be thicker at this point than near the mouth.

After the first drawing operation, the cups are taken to the annealer, again annealed, washed, and dried, in the manner previously described. They are then brought back to the drawing presses for the second drawing operation. After this, they again pass through the operations of annealing, washing, and drying, after which they are taken to the drawing presses for the third or finish-drawing operation, that is performed as previously described.

The transition of the cup A to the case is clearly shown in Fig. 6. Here the three d r a w i n g operations are shown at B, C, and D, respectively. After the three drawing operations, the cases are a g a i n washed and dried (not annealed), when they are taken to the trimmers, one of which is shown in Fig. 7. As shown in Fig. 6, the mouth of the cup is not left perfectly straight after the drawing operations, but has ragged edges. Cracks also sometimes develop in the mouth of the case,

Fig. 7. Automatic Trimming Machine for Trimming the Cases to Length

which make it necessary to trim off a certain amount, to obtain a perfect case.

Trimming the Case to Length. — In the trimming operation, the cases are dumped into the hopper A, shown at the top of the machine, Fig. 7, and pass from the latter into the revolving drum B, to which segments, having pins C driven into them, are fastened. These pins, which are smaller in diameter than the inside of the case, are pointed

and set at an angle. As the segments carrying the pins
rotate, the cases are agitated and drop onto the pins. The
pins now carry the cases to the top of the hopper, and as
they approach the perpendicular position, the cases drop off,
and fall into the chute D, which is connected to the machine
by a close-wound spring E, and tube F. A better idea of
the operation of this trimming machine can be obtained
by referring to Fig. 8. Here the cases are shown dropping
down the chute. They are held by a finger a, which presses

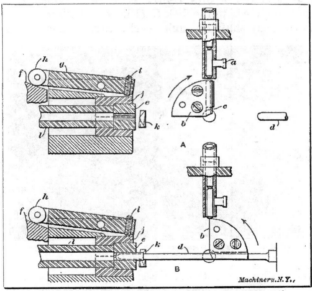

Fig. 8. Diagram showing the Action of the Automatic
Trimming Machine

against them, allowing only one case to drop out at a time.
In the position at A, one case has dropped into the segment
b, and the next case is being held by the finger a. This
finger is released by a cam, located at the rear of the machine. Attached to the face of the segment b is a sheet-
steel plate c, the function of which is to prevent the cases
from dropping out. This sheet-steel plate is held by a
dowel-pin and two screws as shown. Spiral springs are

located under the heads of these screws, thus giving the plate the desired tension on the case.

After the cases are located in the pocket, the segment is revolved in the direction of the arrow, carrying the case into the horizontal position as shown at B. The punch d now advances and carries the case out of the pocket into the chuck e. The chuck begins to close before the punch reaches the end of its travel, so that the punch can force the case in to the correct depth. The punch d is held in a slide, actuated by an eccentric crankshaft which connects the punch-slide G to the disk H (Fig. 7) ; the latter is driven from the rear shaft through bevel gears. When the case is

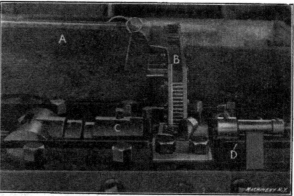

Fig. 9. Horizontal Header with Semi-automatic Feed for Forming the Head on the Case

located in the chuck, the latter is closed by means of a cam located on the rear shaft of the machine, which operates through a lever forcing the beveled sleeve f forward. This beveled sleeve f raises the lever to which the roll h is attached, and closes the chuck by means of the screw i pressing on the outer sleeve j. The punch d then retreats and the trimming tool k advances, trimming the case to length. The inner face of this tool is slightly offset as shown, so that it will take a light shaving cut after cutting the end off. This makes a good finish and does not throw any burrs into the case. The segment b now rotates back in the direction of the arrow as shown at *B*, into the position as shown at *A*, when the cycle of operations is continued. As one case is trimmed it is forced by the following one through the sleeve *l*, which passes through the spindle. The cases pass through this sleeve and drop into a box placed under the machine.

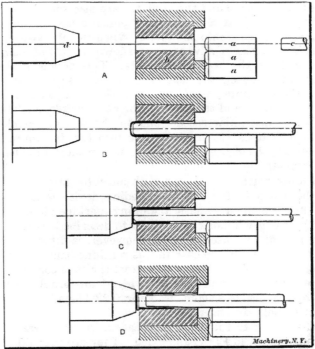

Fig. 10. Diagram showing the Action of Forming the Head on the Case

Forming the Head. — Now that the cases are trimmed to length, they are ready for the heading operation. This is accomplished in a horizontal header of the semi-automatic feed type as shown in Fig. 9. Here the cases are dumped into the hopper *A*, from which they are taken by the operator, who, by means of a shaker, transfers them to the slide B. The cases are placed in this slide with the mouth facing the punch-head C. As the cases come down the slide B, they rest in a pocket, from which they are carried by the punch into the die, where they are headed by the bunter, held in the head D. This heading operation is interesting, and is clearly illustrated in Fig. 10. At A are shown the cases a located in the pocket in front of the heading die b. The heading punch c and the bunter d are also shown back, out of operation. At B the heading punch c has advanced and carried the case into the die b. When in this position, the heading bunter d advances, as shown at C, and commences to form the head on the case. This action of the heading bunter upsets the head of the case, close to the die, so that the heading punch can be withdrawn, as shown exaggerated at D, and the head on the case completed. The punch and bunter now recede, and on the forward stroke of the punch it carries another case into the die, forcing the previously headed one out, which is deposited in a box placed beneath the machine.

A heading machine equipped with an automatic feeding device is shown in Fig. 3. The feeding device is similar to that used on the automatic trimming machine shown in Fig. 7, so that it will not be necessary to describe it further. The feeding of the case to the die, however, is different to that shown in Fig. 9, the case in this machine being fed by two fingers, one of which carries it from the tube connected with the hopper to the other finger which transfers it to the die. This header is seldom used for 0.22 "long," its use being principally for heading 0.22 "short."

Priming. — After the cases are headed, they are washed in the tubs shown in Fig. 2 and dried. They are then taken to the priming department. Here the cases are shaken into plates and a charge of fulminate, which is held in a charger, is inserted in the cases. From the charging room the cases are taken to the priming machines, where they are placed on a friction dial which carries them to a punch. This punch has three grooves filed in its end, the function of which is to distribute the fulminate to the rim of the case. As the punch is kept rotating, it forces the powder to the

rim of the case, by the action of centrifugal force, and
locates it in a manner similar to that shown at G in Fig. 6.
The fulminate is placed in the cases in a wet condition and
will not discharge easily until dry. The cases, after priming, are taken to what is called the "dry-house," where they
are placed in sieves and left until dry in a very warm com-
partment, heated by steam. This completes the operations
on the case.

   Casting the Slugs for the Bullets. — Two principal
methods are used in the manufacture of lead bullets. The
oldest one is to cast the bullets into slugs in a cast-iron mold
of the proper shape, and then swage these slugs to the
required shape in a swaging machine; in the second method,

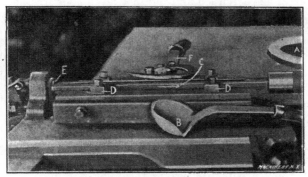

Fig. 11. Molds In which the Slugs for the Bullets are Cast

a lead wire is fed into a special machine which cuts and
forms the bullets to size and shape. Both methods are used
at the present time. The first method is the one employed
by the firm the practice of which is here described. The
second method is described in Chapter VII. When the
bullets are cast into the form of a slug in a cast-iron mold
and afterwards swaged to the proper shape, the slug is
cast in molds made in halves, as shown in Fig. 11. The
molten lead is kept at the required temperature in the pot A,
and is removed from it by the operator with the ladle B.
The lead is poured into the filler C, which is located on top
of the mold by guide blocks D. Then a foot lever situated

beneath the machine is operated, forcing the pin E forward; this pin, in turn, moves the filler in the direction of the arrow, thus shearing the metal which has remained in the filler from that which has run into the mold. The operator now moves lever F to the left, which opens the molds and allows the slugs to drop out. The lever F is then moved to the right, closing the molds, and the operation is continued.

This operation is more clearly illustrated in Fig. 12, where a section of the filler and mold are shown. At A the filler, a is shown in line with the holes in the mold b, when the metal is being poured in, and at B the filler is in the position that it occupies in relation to the mold after the metal has solidified and the foot lever is depressed. The

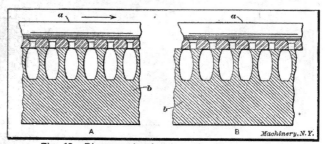

Fig. 12. Diagram showing how the Slugs are Cast

surplus metal which is left in the filler is tapped out into the pot and remelted after the slugs which stick to the filler have been scraped off.

Tumbling and Inspecting the Slugs.— The slugs as they come from the molds have fins and burrs on them, due to several reasons; one is that the operator does not close the mold tightly; another, that dirt or scrap gets in between the two halves of the mold. If the slugs were taken to the swager in this condition, they would not pass down the tube, so it is necessary to tumble them to remove the fins and burrs. This is done in an ordinary tumbling barrel which is revolved slowly. After the slugs are tumbled, they are then dumped on a bench and inspected. In this inspection, "half-slugs" and imperfectly formed slugs are removed.

"Half-slugs" are due to the molder not having sufficient metal in his ladle to fill the mold. These would make imperfectly formed bullets of light weight.

Swaging the Bullets. — After the slugs have been inspected and all half-slugs or imperfectly formed ones have been picked out, they are dumped into a hopper *A* (only the lower portion of which is shown), located at the top of the swaging machine shown in Fig. 13. From this hopper they

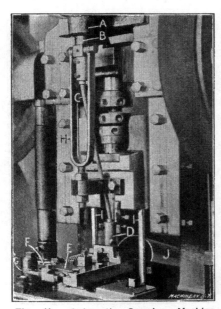

drop down a tube *B* into the close-wound spring *C*. This spring connects the tube *B* with the pocket or receptacle *D*, located over the finger-slide *E*. From the pocket the slugs are carried to the dies by fingers, which are held to the slide *E*. This slide is actuated by a bell-crank *G*, which is given a reciprocating motion by a cam *F*, fastened to a vertical shaft, and driven from the crankshaft through bevel gears. The slugs would not come down the sleeve and spring of their

Fig. 13. Automatic Swaging Machine which forms the Slug into a Bullet

own accord, so it is necessary to agitate them. This is accomplished by fastening a yoke *H* to the ram of the press, and attaching this yoke to the sleeve *B*. The movement of the ram carries the sleeve *B* up and down in the hopper, which action agitates the slugs and causes them to drop down. The bullets are removed from the die by a knockout connected to the ram of the machine by two studs *I*.

The action of swaging the bullets is more clearly shown in Fig. 15. The swaging dies are made in two pieces and are ground and lapped on the surfaces which come in contact. At A the slug a is shown as it drops down into the die b and is located on the die-pin c. In the position shown at B the ram of the press has descended, carrying the punch d into the die, which action forms the bullet. The punch, in forming the bullet, forces the excess material out of the vent hole provided in the upper die. This action is very interesting, as the excess material is gathered from the slug and forced out of the vent hole in the form of short wire. As the bullet is formed, the ram of the press again ascends and in its ascension the die-pin c is pushed up through the dies, carrying the base of the bullet flush with the top of the die. The bullet is removed from the top of the die by the fingers as they carry another slug to the die, and falls into the chute J, shown in

Fig. 14. Semi-automatic Loading Machine for Seating the Bullets in the Cases

Fig. 13. This swaging operation finishes the bullet to the exact size and also to the correct weight. The bullet for the 0.22 long-regular weighs thirty-five grains.

Loading. — Now that the case and bullet are completed, they are ready for loading. Both the bullets and cases are removed to an outside building where the loading machines are located. The cases or shells are first shaken into what is called a "shell plate," which has a baseplate doweled to it. The bullets are shaken into a bullet plate, and the powder is then put in a charger which has holes in it registering with the holes in the shell plate, and slightly smaller in diameter than the inside of the cases. The thickness of these charging plates governs the amount of powder that is put in the cases. The charging plate is now located over the shell plate and tapped slightly, causing the powder to drop into the cases. The charger plate is then removed and the bullet plate substituted, being located by dowels. Both plates are now taken to the semi-automatic loader, Fig. 14. (This loader is shown with the plates removed). The plates are put on the table A, and held by the clamp-bolt B. When everything is set properly, the operator

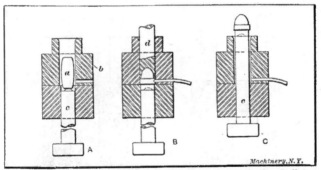

Fig. 15.   Diagram showing the Action of Swaging the Bullets

presses the hand lever C, thus starting the machine. He then steps back from the machine, as occasionally a number of cartridges, and, in some cases, a whole plate of cartridges, explode, making it dangerous for him to stand in close proximity to the press.

The table A on which the loading plates are held is moved forward by means of a pawl engaging in a rack, fastened to the under side of the table. The table is moved a distance equal to the space between a row of holes for each stroke of the press. The pawl is actuated through a series of levers and the arm D which is connected eccentrically to the crankshaft of the press. When the last row of cases in the plate has been operated on by the row of punches E, the machine is automatically stopped by a trip lever at the back of the machine.

A clearer idea of the method of loading the cartridges may be obtained by referring to Fig. 16. Here at *A*, the cases and bullets are shown located in the shell and bullet plates *a* and *b*, respectively, ready for assembling, or seating; and at *B* the bullets and cases are shown assembled, by the action of the seating punches *c*. The plates are now removed from the press, and the slip plate *d* removed, when the loaded cartridges drop out. After the bullets have been seated in the cases, the loaded cartridges are taken to an automatic machine, where they are crimped and cannelured.

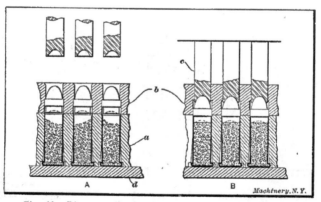

Fig. 16. Diagram showing how the Bullets and Cases are
Assembled

Crimping and Canneluring. — Crimping the cartridges consists in tightening the case around the bullet, as shown at *I* in Fig. 6, to prevent the latter from falling out. This operation is performed in the automatic machine shown in Fig. 17. The loaded cartridges are dumped into the hopper *A* through which passes a belt (inclosed in the box *B*) having scoops fastened to it. Theses scoops *C* carry the cartridges out of the hopper up to the top of the slide *D*. Here the cartridges drop out of the scoops into the slide. The slot in this slide is slightly larger than the body of the case, but is smaller than the head, so that it is impossible for the cases to go down the slide unless they are head upwards. As the cartridges come down the slide D, they come in contact with the wheel E, which is rotated by a round belt F. This wheel has slots cut in it, in which the cartridges hang, the under sides of the heads bearing on the periphery of the wheel, and as the latter is kept rotating it deposits the cases on the revolving dial G. The guard L prevents the cartridges from dropping out before they reach the dial. As the cartridges drop out of the wheel E onto the dial G, they are guided by the block H. The dial G rotates in the direction of the arrow and carries the cart-

Fig. 17. Automatic Crimping and Canneluring Machine

ridges around where they are lined up by the guide I. As they continue in their travel, they pass between the stationary segment block J and the revolving dial K, the action of which rotates the cartridges and performs the crimping and canneluring operations. As they pass around still further, they are removed from the dial by a guide and drop into a box. The canneluring is done by means of narrow knurled projections formed on the edges of the dial K and the segment block J, dial K being fastened to the dial G. The crimping is also done by forms on the above-mentioned parts.

Greasing, Packing, and Testing. — The cartridges are now completed, as regards the manufacturing operations, and are ready for greasing and packing. Greasing is only done when the bullets are made from commercially pure lead, and not of composition of from 3 to 5 per cent tin. When made of this composition, it is not necessary to grease the bullets. The object of the greasing is to prevent them from leading the bore of the rifle. The 3 to 5 per cent composition has the same effect as the grease, but is much cleaner and of a more finished appearance. The canneluring, as shown at *I* in Fig. 6, forms narrow knurled grooves during the crimping operation for the purpose of holding the grease. If the bullets are made from commercially pure lead, they are shaken into plates and dipped into molten grease. The grease just sticks in the canneluring, leaving the remainder of the bullet practically clean—that is, if the grease is at the proper temperature. If slightly cooler than the correct temperature, the grease will form in clogs on the bullet, which have to be removed by wiping them with a rag. The shaker plates are now located on a packing plate. This packing plate has three hundred holes drilled through it in groups of fifty, which is the number of cartridges put in each box. The shaker plate holds only half the number of cartridges that the packing plate does, so that it requires two shaker plates to fill the packing plate.

In operation, the first shaker plate is placed over the packing plate, and both plates are then turned over, when the shells drop into the packing plate. The shaker plate is then removed and a slip plate substituted for it. Another shaker plate is placed over the packing plate and the same operation repeated. This fills the packing plate and leaves half of the cartridges in the plate with their heads up, while the other half are reversed. The cartridges are now removed from the packing plate and placed in a slide provided with compartments, from which they are removed and placed in boxes. This packing arrangement is semiautomatic, and is considerably quicker than packing by hand.

Even though the cartridges have passed by all the inspections necessary for the manufacturing operations, they are still not ready to ship, but have to pass through a rigid test. A number, picked at random, are taken to the testing department where the tester in charge tests them for accuracy. There are various methods of testing the accuracy of the cartridge. One is to locate the rifle in a stock which holds it rigidly. The sight is then located accurately over the bull's eye, the trigger pulled, and the result noted. The cartridge is also tested for accuracy by off-hand shooting, and other similar tests.

CHAPTER IV

MANUFACTURE OF CENTER-FIRE CARTRIDGES

THE greater number of military cartridges used at the present time by the various governments are about 0.30 caliber. This size is also used for many sporting rifle cartridges because of the high velocity and penetrating power and the comparatively flat trajectory that can be obtained. In the present chapter, the practice of the Dominion Cartridge Co. in the manufacture of 0.30-30 Winchester cartridges is described.

Chief Requirements of a High-velocity Rifle Cartridge. — Among the chief factors governing the accuracy with which a bullet strikes a target may be mentioned: 1. The rifling in the barrel; 2. The nature and the amount of charge behind the missile; 3. The shape and equilibrium of the bullet. The rifling and the explosive charge behind the bullet have a direct effect on the velocity, and also govern the height of the trajectory curve. Theoretically, it is impossible for a bullet to travel through the air in a straight path, but with the high-powered rifles, considering a range of one hundred yards, the height of the trajectory curve at fifty yards is very slight. In the case of the cartridge to be described, for example, the height of the trajectory curve at fifty yards is only 1.280 inch. The shape of the bullet has not such a pronounced effect on its accuracy as has its equilibrium—that is to say, if a bullet does not balance properly on its axis, it is impossible for it to travel in a straight path. This is one of the problems a manufacturer of cartridges has to deal with, and is, in all probability, one of the most difficult to master.

Probably the best-known American high-powered rifles using smokeless powders are the 0.30-30 Springfield, Winchester, Marlin, and Savage rifles. The bullet for the cartridges used in these rifles is made in three distinct patterns, viz., the full metal-cased or hard-point, the part metal-cased or soft-point, and the mushroom or hollow-point. The

41

first-named of these is used more particularly for military purposes, while the two latter types are used for hunting and sporting purposes in general. The soft-point bullet which weighs about 170 grains is the type most commonly used for hunting purposes. When fired, this bullet has a muzzle velocity of approximately 2000 feet per second—a rate of 23 miles per minute.

First Operations. — The case for the 0.30-30 cartridge is received in the form of a cup, as shown at *A* in Figs. 1 and 2. These cups are taken to the annealer, shown in Fig. 1,

Fig. 1. Evolution from the Cup to the Finished Cartridge

Chapter III, where they are annealed, washed, and dried, as previously described. When dry, the cups are taken to the friction-dial drawing press shown in Fig. 3, where they are placed on the stationary table *A*, from which they are removed by means of a shaker to the revolving dial *B*. This drawing press is similar to that shown in Fig. 4, Chapter III, except that it is single-headed. After the cup has passed through two drawing operations, as shown at *B* and *C* in Figs. 1 and 2, it is ready for the inserting or indenting operation.

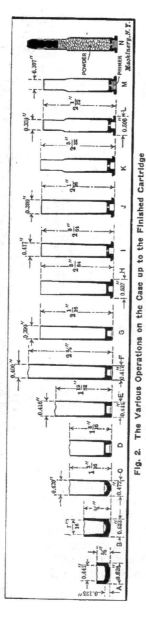

Fig. 2. The Various Operations on the Case up to the Finished Cartridge

Rough-forming the Pocket for the Primer. — The case for the 0.30-30 Winchester cartridge is made with a solid head, which is not the case with cartridges of smaller size, in which, as a rule, smokeless powders are not used. The solid head is necessary to withstand the high pressure developed by smokeless powders. After the second drawing operation, the cup is not annealed, but is taken directly from the drawing presses to the headers, one of which is shown in Fig. 4. The header is used for inserting the pocket as well as forming the head. The principle on which this header works is similar to that of the horizontal header shown

in Fig. 9, Chapter III. The method used in rough-forming the pocket is clearly shown at A in Fig. 5. As the case comes down the slide of the header shown in Fig. 4, it is located in a pocket, from which it is carried by the punch a, Fig. 5, into the die b. Here it is held by the punch while the inserting bunter advances and forms the pocket. Both punch and bunter then retreat, and, on the forward stroke of the ram carrying the punch, the case previously inserted is forced out of the die, as the punch carries another case in. The die b is made a good fit for the case, which, when pushed through, expands and is stripped from the punch.

43

Third and Fourth Drawing Operations. — The case is not annealed after the second drawing operation, and it is necessary to do this before it can pass through the successive drawing operations. The cases are taken to the annealer, annealed, washed, and dried, and brought back to the drawing presses. The cases are now of considerable length, and it is not advisable to perform the third and fourth drawing

Fig. 3.   Friction-dial Drawing Press for Drawing the Cups

operations in the press shown in Fig. 3, as they would not stand up properly on the dial. For these operations, a ratchet-dial drawing press is used, the design of the dial of which is somewhat similar to the dial of the swaging machine shown in Fig. 16. After the third drawing operation the cases are again annealed, washed, and dried, and then pass through a fourth drawing operation. (*F*, Fig. 2.)

Trimming the Cases to Length. — As shown at *F* in Fig. 2, the top edge of the case is extremely ragged. Cracks develop in the mouth of the case which make it necessary to remove a certain amount to obtain a good finish. This is accomplished in the semi-automatic trimming machine shown in Fig. 6. Here the cases are placed in the slide *A* by the operator, from which they pass into a pocket at the base of the slide. From the pocket they are carried into the chuck by means of the punch *B* held in the punch-head *C*, which is actuated by an eccentric crankshaft. The chuck begins to close before the punch reaches the limit of its

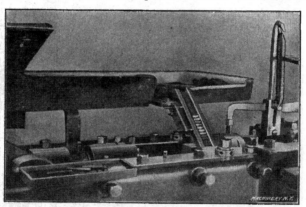

Fig. 4.   Horizontal Header of the Semi-automatic Feed Type
for Inserting the Cup and Forming the Head on the Case

travel, thus allowing the punch to insert the case to the desired depth. The chuck is closed by means of a cam (not shown) at the rear of the machine, operating a clutch, which, in turn, forces a sleeve forward, thus closing the chuck. The tool-slide *D* carrying the trimming tool *E*, is now brought forward, and trims the case to the desired length. As the trimming tool *E* advances, the punch *B* retreats from the case. The chuck is now opened, and the punch advances carrying in another case, which forces the previously trimmed one into a hollow sleeve. This hollow sleeve passes through the spindle of the machine, the trimmed cases dropping through it into a box placed beneath the machine. The pulley F drives the camshaft, which, in turn, operates the chuck-closing, trimming, and case-inserting mechanisms. The machine is started by operating the lever G.

Forming the Head, Pocket Sizing and Piercing. — The case is now of the shape shown at G in Fig. 2, and is ready for the heading and stamping operation. It is again taken to the heading machine shown in Fig. 4, and operated on as shown diagrammatically at B in Fig. 5. As before, the case is placed in the slide and drops down into the pocket,

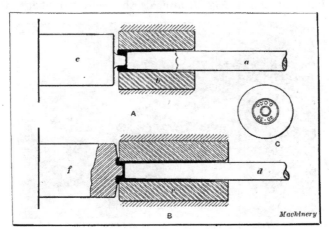

Fig. 5. Illustration showing how the Cup is Inserted and the Case Headed

from which it is carried by the punch d into the die e. The heading bunter f now advances, finishing the pocket, and expanding the end of the case to form the head. In this operation the punch does not retreat, but remains in position, supporting the case, while the head is being completed. An end view of the heading bunter is shown at C, and the shape of the case after the heading operation is shown at H, Fig. 2. There is considerable wear on the teat of the heading bunter in the heading operation, thus making it necessary to size the pocket so that the primer can be inserted without undue pressure. For this purpose, the cases are taken to a sizing machine of the ratchet-dial type, where they are placed on pins driven into this dial, and pass successively under a piercing and a pocket-forming punch. The pierced case is not shown in Fig. 1, but may be seen at *I* in Fig. 2, where all the other operations on the case are also more clearly shown.

Mouth Annealing. — When the pocket has been sized and the hole pierced in the case, thus making an opening so that the powder can be exploded by the discharge of the primer, the cases are taken to a "mouth annealer." This machine, which is shown in Fig. 7, anneals the cases for about two-

Fig. 6. Trimming Machine of the Semi-automatic Feed Type for Trimming the Cases to Length

thirds of their length. The cases are placed in a vertical position, resting on their heads on the revolving dial *A*, which is rotating in the direction of the arrow. They pass around on this dial between the guard *B* and the spring *C*. This spring is given a vibrating motion by the action of the revolving dial, thus agitating the cases and arranging them in single file, so that each case will be exposed to the flame. As the cases are carried around on this dial they pass in front of two gas burners *D* and *E*, where the mouth of the case is annealed. Gasoline is used as a fuel, being pumped into the burners at the desired pressure by a pump located at the right of the annealer, but which is not shown in the illustration. The speed at which this dial revolves is such that the cases remain in front of the burners long enough to be sufficiently annealed. They are then removed from the dial by a wire *F* which pushes them off into a box, where they are allowed to cool gradually.

Reducing. — The reason for annealing the mouth of the case is to make it soft, so that it can be reduced on the mouth without cracking or folding. Before the cases are reduced, they are oiled with a rag which has lard oil spread over it. They are then taken to the machine shown in Fig. 8, which is a reducing press of the ratchet-dial type. Here they are

Fig. 7. Friction-dial Mouth Annealer

dumped into a box placed in front of the machine, from which they are removed by the operator and placed in the holes in the ratchet dial *A*. This dial is driven by a finger *B*, which is held to a dovetailed slide *C*, the slide being actuated by the lever *F*, which is connected eccentrically to the crankshaft. The holes in the dial *A* are larger at the front end. The dial is made in this manner so that the heads can be inserted in the larger hole, and, as the dial revolves, the friction between the head of the case and the base of the machine draws the case back into the smaller portion of the hole. When the case is in this position, it cannot be removed by the reducing dies, should they stick to the case.

The friction between the head of the case and the bed of the machine, however, cannot be relied upon to locate the cases properly, so a spring pad is placed in the bed of the press, over which the cases pass before reaching the first reducing die.

The ram of this machine is made to hold two reducing dies. The first boss D holds what is called the breaking-down die, which only passes down a certain distance over the mouth of the case, while the second boss E holds the reducing die. This latter die travels down practically the whole length of the case, and gives it a tapering shape. The action of reducing is more clearly shown in Fig. 9, where the dies are located in the relative positions that they occupy when in the machine. The breaking-down die is shown at A while the reducing die is shown at B. It is necessary in this operation to support the inside of the case while reducing, and for this pur-

Fig. 8. Ratchet-dial Reducing Press for Reducing the Mouth of the Case

pose punches a and b are inserted in the die, as shown, to prevent the case from folding.

Verifying. — From the reducing machine the cases are transferred to the verifying machine shown in Fig. 10. Here they are placed by the operator in verifying dies, sixteen of which are held in the ratchet dial A by spanner nuts B. This ratchet dial A is actuated in the same manner as that of the reducing press. As the cases pass around, the punch C seats them in the verifying dies, and as they pass around still further, the punch D forces them down into the die until the under side of the head of the case rests on top of it. As the ratchet dial continues in its travel, a "knockout" placed beneath the dial lifts the ejector pins, which forces the case up out of the die to a sufficient height, so that the pick-up E can grip it. As each case is picked up, it forces the preceding one up through a brass tube from which it falls out at F into the chute G. This pick-up is operated by a lever H fulcrumed to the ram of the machine and pivoted to a bracket I, which is fastened

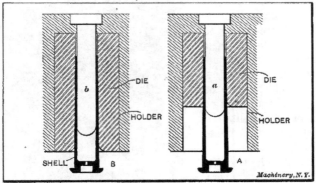

Fig. 9.   Showing how the Mouth of the Case is Reduced

to the uprights of the machine. The end of the lever which operates the pick-up is rounded, so that it "rolls" freely between the projections J formed on the pick-up spindle K. This verifying operation reduces the cases on the mouth to the correct diameter, as shown at K, Fig. 2. Small pins or knockouts are used to support the mouth of the case while being reduced, acting on the same principle as those shown in Fig. 9.

Trimming. — After the cases have been verified, they are removed to the washing-room, where they are put into the revolving tubs shown in Fig. 2, Chapter III, washed, dried, and then brought to the trimming department. The operations now to be performed on the case are trimming the head, trimming the case to the proper length, and burring the mouth so that the bullet can be easily inserted. The trimming machine for performing these operations is shown in Fig. 11. The cases are placed in the slide A by the operator, from which they pass into a pocket B, the head of the case facing the punch C, as shown. The drum D to which cams are attached actuates the slide E, carrying the punch C, which, in turn, forces the case into the revolving chuck F. This chuck is made in two pieces, and is closed by split rings F_1, operated upon by a cam, attached to the main driving shaft beneath the machine. The chuck is rotated by a belt G, which runs on the pulley H. Now that the cases are held in the chuck, a cam attached to the driving shaft carries a slide I forward, in which is held a trimming-tool holder J_1. Circular forming tools are held in this toolholder for trimming

Fig. 10. Verifying Machine in which the Reducing of the Case is Completed

the head of the case, giving it the appearance as at L, Fig. 2. A group of these circular tools may be seen hanging on the hopper at J, to the left of the illustration. At the same time that the slide carrying the head-trimming tool is being advanced, the slide K also advances, carrying a trimming tool, which trims the case to length. This trimming tool is similar to that shown in Fig. 8, Chapter III. As soon as the case is trimmed to length, the mouth-trimming tool *L* advances and trims the mouth of the case. This tool is made from ⅜-inch drill rod.

After the trimming operations are completed, the chuck is automatically opened, and the punch *M* advances carrying the case out of the chuck, from which it drops into the chute *N*, and thence into a box *O*. The snap-gage used for gaging the head and length of the case in this operation is shown at *P*. It is of the ordinary type of combination snap- and ring-gage. Gage *P* is only used when setting-up the machine, and testing at the beginning of each box of cases. After the cases come from this machine, they are

Fig. 11. Trimming Machine for Trimming the Head and Mouth, and for Trimming the Case to the Desired Length

deposited on a bench, where an operator passes them through a snap-gage shown in Fig. 12, which is attached to the bench. This gage is so constructed that it is impossible to pass a case through the hole *A* and into the box, without first passing the head of the case through the slide of the gage. If any cases are found to have large heads, they are put to one side and again pass through the trimming operation, so that all cases that pass this inspection have heads of the correct diameter.

Inserting the Primers and Inspecting. — The cases are now transferred from the trimming department to the priming machines, one of which is shown in Fig. 13, where they are placed in hollow pins *A*, twelve of which are driven into a dial *B*, fastened to a ratchet dial *C*. This ratchet dial is driven in a manner similar to the other ratchet dials previously described. As the operator places the cases in the hollow pins *A*, they pass around under a punch (not shown) which is held in the boss *D*. This acts as an emergency punch to insure that all the cases have been pierced. A plate *E* acts as a guard to prevent the emergency punch from pulling the cases off the pins. However, this plate *E* is not relied upon exclusively, a punch *F* which is held in the ram of the press being used for seating the cases prop-

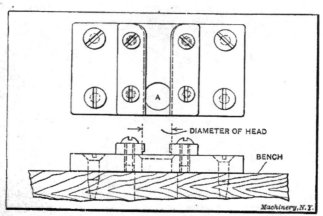

Fig. 12. Snap-gage used on the Bench for Gaging the
Head of the Case

erly in the hollow pins. The primers are shaken into small vulcanite plates, from which they are transferred to the table *G*, located at the rear of the machine. A friction dial, which is operated by a round belt driven from the main crankshaft, rotates just in front of this table. The operator now shoves the primers from the table onto the friction dial, which carries them around between two guards. This action of feeding the primers is similar to the action of the friction-dial drawing press, shown in Fig. 3. A finger *H* held on a slide carries one primer at a time out from the dial, and holds it central with the pocket in the case. The

punch *I* now descends, and carries the primer out of the
finger, seating it in the pocket. As the dial passes around
still further, the pick-up *J* descends, lifting the cases from
the pins, and transfers them to a box in a manner similar
to that shown in Fig. 10.

As can be seen in the illustration, the pins *A* are tapered
so that the pick-up jaws can pass down over the head and
get a good grip on the case. The reason for using hollow
pins instead of the ordinary solid pins for
this operation is that
the mouth of the case
is considerably smaller than the upper part
of the body. This
makes it necessary to
put the cases in hollow pins, so that they
will not be able to
"wobble" when the
primer is inserted, as
would be the case if
solid pins were used.
After the primers
have been inserted in
the pockets of the
cases, they are transferred to a bench
where an operator, by
means of a small
straightedge which
is rubbed across the top of the case, tests the primers to
see whether any of them project above the face of the
head. If a primer projects, a bright spot is noticed; this
primer must be knocked out and another one inserted.

Fig. 13. Priming Machine for Inserting
the Primer in the Pocket of the Case

Slugs for Bullets.— The bullet referred to in the following is of the so-called "soft-point" type, and consists of a
lead center which is partially enveloped by a metal case.
The soft-point bullet is also known as the "mushroom" bul-

let, and is used chiefly for sporting purposes. For military purposes, a cupro-nickel or steel case completely envelops the lead bullet, so that there is no mushrooming effect when an object is penetrated. This completely enveloped lead bullet when pointed is known as a "spitzer" bullet. The manufacture of this type is described in Chapter VII. The various stages through which the soft-nose bullet passes during its manufacture are illustrated in Fig. 14, where A is the slug as obtained from the molds. These slugs are cast in molds of a design similar to those shown in Fig. 11, Chapter III, but, of course, the forms in the molds are of the desired shape. The slugs are removed from the molding department to a tumbling barrel, where they are tumbled for a considerable length of time to remove the fins;

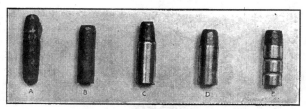

Fig. 14. Operations performed on the Lead Filling before and after Assembling in the Metal Case

then they are dumped into the hopper of an automatic swaging machine, similar to that shown in Fig. 13, Chapter III, and come out in the form shown at B, Fig. 14. This finishes the operation on the lead center.

Drawing and Trimming the Metal Case. — The metal case, which is shown assembled on the lead center at C, D, and E in Fig. 14, is made from copper and is nickel-plated after it has been trimmed to the desired length. This case for the bullet comes in the form of a cup and passes through three drawing operations, and the annealing, washing, and drying operations, as described in Chapter III, regarding the shell for the 0.22 "long." The same class of machines is also used for the various operations on this case.

When the case is completed, that is, after the drawing and trimming operations are performed, the case and the lead

centers are transferred to the loading department. Here the
lead centers are shaken into one plate while the copper cases
are shaken into another. These plates are now assembled
over each other, the plate holding the cases being on top,
and a slip-plate is put under them. The three plates are
then placed in the loading machine, Fig. 15. Seating
punches *A* are held to the ram of this machine and, as the
lever *B* is pulled down, the ram descends, and the seating

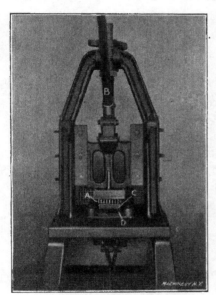

Fig. 15.   Hand Loading Machine for
Loading the Cartridge

punches f o r c e t h e
cases out of the plate
C over the lead cen-
ters held in the plate
D.   T h e s e seating
punches are made so
that they pass into the
holes in the shell plate.

The bullet plate or
"center holder" is
made to give the bul-
let the shape shown
at *C*, Fig. 14.   The
bullets as now form-
ed are again transfer-
red to the swaging de-
partment, where the
operator places them
on the table *A* of the
machine shown in Fig.
16, from which they
are put in the ratchet
dial *B*, the lead point

extending downward.   As the dial passes around, the
punch *C* seats the bullet properly in it, and as it continues in its rotary movement, the punch *D* forces the
bullet out of the dial into a die.   This die gives the
bullet the shape shown at *D*, Fig. 14, and also makes it
symmetrical, so that it will balance properly on its axis. It
might here be mentioned that the finishing of the bullet in
one die insures this result.   As the ram of the press ascends,

it carries upward two rods connected with the knockout motion under the press, which, in turn, force a punch up through the die, the action of which again transfers the bullet from the die to the dial B. Just as the bullet is located in the dial, the latter is revolved and the punch E forces the bullet out of the dial into a chute, from which it is deposited into a box.

Forming Knurled Grooves. — The next operation on the bullet is the forming of the knurled grooves shown at E, Fig. 14. The knurling of these g r o o v e s is accomplished in the canneluring machine shown in Fig. 17. The bullets are dumped on the dial A, from which they are removed by the o p e r a t o r and placed in a vertical position on the revolving dial B. Attached to the revolving dial B is a dial C, which has two knurled projections on its periphery. Back of dial C is a segment D, which also has two knurled projections on its face.

Fig. 16.   Swaging Machine for Forming the Lead Filling and the Metal Case to the Desired Shape

As the bullets pass around on the dial B in the direction of the arrow, they are rotated by the action of the dial C revolving against the stationary segment D. This action forms the knurled grooves entirely around their peripheries. As the dial carries the bullets around still further, they are removed from it by means of a guide E, and are dropped into a box placed under the machine. The object of this canneluring operation is to form a groove in the bullet so that the top of the case can be turned in, thus holding the bullet more securely in the case. After the canneluring operation, the bullets are taken to the swaging machine in Fig. 16, and again pass through the operation of swaging, the same die being used as before. This is to correct any eccentricity of the bullet which might have been caused by canneluring.

Loading. — The case and bullet are now ready for assembling or loading, and are transferred to one of the explosive departments where this operation is accomplished. The cases are put into one plate and the bullets into another.

Fig. 17. Canneluring Machine for Forming the Grooves
in the Bullet

Then the desired amount of powder is shaken into a charger, which is located over the shell plate and rapped slightly, thus depositing the powder in the cases. The shell plate containing the cases and the bullet plate are removed to the loading machine shown in Fig. 15. Here the shell plate is put into the slide, the bullet plate located on top of it by means of dowels, and the handle B pulled down, carrying down the ram of the press to which the seating punches are fastened. These seating punches force the bullets out of the bullet plate and locate them in the cases. The plates are then removed and the loaded cartridges dumped out. A different shell plate is used in this operation.

Crimping. — The loaded cartridges are now transferred to the crimping machine, Fig. 18, in which the top of the case is tightened around the bullet. The operator dumps the cartridges into a box, which is held on the brackets A in front of the machine. He then removes the cartridges from the box and places them in the holes in the dial B. This dial rotates to the left, and as the cartridges come below the punch C, the bullet is seated in the case to the correct depth. Two bosses E are provided on the ram of this machine. Both bosses, however, are not used for crimping, as this machine is also used as a reducing press, when both bosses are necessary. The crimping die can be held in either boss, but is held preferably in the one to the left. This die is made so that it passes over the bullet and turns the case in the groove, thus tightening the case securely on the bullet. As the cases pass around still further, they are removed from the dial

Fig. 18. Crimping Machine for Tightening the Cartridge Case on the Bullet after Loading

by the wire F and are deposited in a box under the machine. Spring pads are also used under this dial, the purpose of which was explained in connection with the reducing press shown in Fig. 8.

Testing and Packing. — The cartridges are now finished as regards the manufacturing operations, but they are not ready for the market until they pass through a rigid inspection. This inspection consists in testing for accuracy, velocity, and penetration. The accuracy of the cartridge is tested by means of shooting the bullets at various ranges for which the cartridges are adapted. This is done both by off-hand shooting, and also by locating the rifle in stocks, sighting it directly over the bull's-eye, firing it, and then noting the results. The velocity of the cartridge is determined by means of an instrument called the chronograph. This is an electrical instrument which operates in the following manner: The recording mechanism is connected with a wire, which is held just in front of the muzzle of the gun from which the cartridge is to be fired. Another wire is placed in front of the target and connected to the chronograph by an electric circuit. When the apparatus is adjusted, a signal is given for the cartridge to be fired. As the bullet leaves the muzzle of the gun, it cuts the first wire which is connected to the chronograph, and the latter commences to record the flight of the bullet. When the bullet strikes the target, it breaks the electric circuit connected with the chronograph, and the instrument instantly stops registering. The recording apparatus shows the time taken by the bullet in traveling from the gun to the target, and as this distance is always known, it is an easy matter to determine the velocity in feet per second.

The cartridge is tested for penetration by shooting the bullets into boards of a given thickness. Pine boards, 7/8-inch thick, are, as a rule, used for this purpose. The penetration of the 0.30-30 soft-nose bullet, which is described in this chapter, is eleven boards, at a distance of 15 feet. If the cartridge does not pass this inspection satisfactorily, the cause is ascertained and rectified before any are shipped. These cartridges are packed by hand in boxes which hold twenty-five. The boxes are then put in cases holding 5000 cartridges.

CHAPTER V

FRANKFORD ARSENAL METHOD OF DRAWING CARTRIDGE CASES

THE two preceding chapters describe the methods used by the Dominion Cartridge Co. in producing metallic cartridges. The present chapter deals with the methods followed by the Frankford Arsenal, Philadelphia, Pa., in draw-

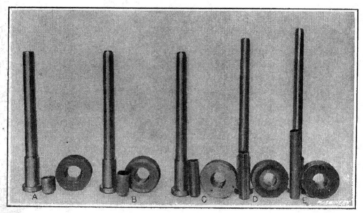

Fig. 1. Sequence of Redrawing Operations on a 0.30-caliber Cartridge Case, and Tools used

ing the cases for 0.30-caliber cartridges. In the following chapter, the tools used for the drawing operations will also be described in detail.

Making the Cups for Cartridge Cases.— The first operation in the making of a cartridge case is to produce a cup, then by successive redrawing operations this cup is reduced in diameter and extended to the required length. Fig. 1 shows the various steps in the sequence of redrawing operations following that of making the cup. From this illustration it can be seen that five operations are necessary to bring the case to the required length—these are called redrawing operations because the work accomplished consists in reducing and redrawing a piece that has already been drawn to cup form. The press used for making the cups from which cartridge cases are made is shown in Fig. 2. It is of the double-action type, and carries four punches and dies, thus making four cups at each stroke. This punch press operates at 100 revolutions per minute, producing 400 cups a minute, or 24,000 cups per hour. The type of blanking and cupping dies used in this machine are shown diagrammatically in Fig. 3, which gives a comprehensive idea of the action that takes place in the formation of the cup.

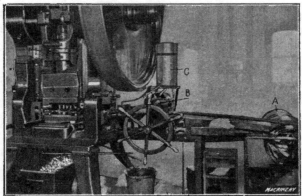

Fig. 2.   Making the Cups, Four at one Stroke of the Press, at the Rate of 24,000 per Hour

At *A* the blanking punch is shown in contact with the top face of the brass sheet. At *B* the blanking punch has cut out a disk of the required size and carried it down to the first shoulder in the combination blanking, cupping, and drawing die. At *C* the combination cupping and drawing punch has come into operation and has started to form the blank to cup shape; and at *D* the blank has been forced completely through the die and has been given the first drawing operation.

The sheet stock is held on the roll *A* located to the right of the machine shown in Fig. 2, and is drawn into the press under the blanking and drawing punches by means of feed-

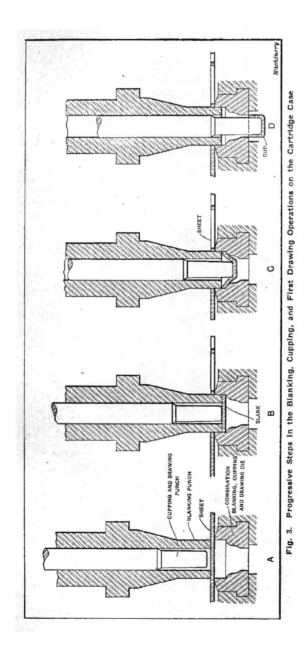

Fig. 3. Progressive Steps in the Blanking, Cupping, and First Drawing Operations on the Cartridge Case

rolls. After the stock passes through the rolls B, it is oiled by means of a rag saturated with lard oil that is contained in the tank C. The sheet is fed a distance equal to the diameter of the blank plus the width of the web for each stroke of the press (see Fig. 4), and after it has been started by the operator, who uses the handwheel feed shown, it is carried on automatically by feed-rolls located at

63

both ends of the throat of the press. As the sheet from which the blanks have been sheared protrudes from the left-hand end of the machine, it passes through a shearing die, which, in conjunction with a knife operated by a crank and connecting lever held on the extreme end of the crankshaft, cuts off the scrap, enabling it to be packed in boxes. The chief reason for cutting up the stock in this manner is to avoid having it pile up around the machine and also to enable it to be more easily removed.

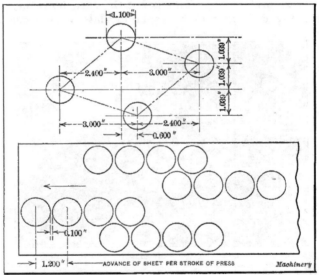

Fig. 4. Manner of laying out Combination Blanking, Cupping, and Drawing Dies in Order to Economize in Stock

The manner in which the dies and punches are laid out in order to economize in stock is shown in Fig. 4. By.referring to the upper diagram in this illustration, it will be seen that the centers of the four dies are located in a "diamond" shape, thus reducing the width of the sheet required and securing the most satisfactory lay-out for the punches and dies. The condition of the sheet after the press has made four strokes is shown in the lower portion of the illustration, which indicates the process in cutting out the blanks.

Setting Drawing Punches and Dies. — To set drawing punches and dies properly requires considerable experience, as this is a difficult task under the most favorable conditions. The dies and punches are usually aligned with each other by setting them in the approximately correct position, then running through a few cups and noting the results. In the type of die-holder shown in Fig. 5, it is not necessary, of course, to reset the dies when they have been removed for grinding, if proper attention has been paid when they were first set up. This is not the case, however, with redrawing dies, as will be explained later. One peculiar

Fig. 5. A Closer View of the Machine shown in Fig. 2

point in making cups that causes considerable trouble is that it is practically impossible to produce a cup with a straight top; that is, one in which the metal is drawn to the same extent on one side as it is on the other. The reason for a cup drawing irregularly in this manner is not due in all cases to inaccurate setting of the punches and dies, but generally to a variation in the thickness of the sheet from which the blank is cut out. It is a peculiar fact, but nevertheless true, that it is practically impossible to roll sheet metal uniform in thickness; that is to say, the sheet is thicker in the center than it is at the outer edges. The reason given for this is that the rolls, even though they are 8 or 10 inches in diameter, spring to a slight extent in the center—where they are unsupported—and thus produce a sheet of varying thickness throughout its width. Another difficulty experienced in making cups is the striking of hard

Fig. 6. Drawing Press used for First, Second, and Third Redrawing Operations—Four Cups per stroke at the Rate of 24,000 per Hour

spots in the metal. It is obvious, of course, that if the stock is not of a uniform hardness, the softest spot or portion will draw much more than the harder portion, and, hence, a cup having one side longer than the other will be obtained. Not only will the top edge be irregular, but the walls will also be of uneven thickness. It is claimed by those who have had experience in this work that it is impossible to rectify any defect of this kind in the succeeding redrawing operations. When a cup is once started with a wall of unequal thickness, this condition prevails until the final drawing operation, so that it will easily be seen that great care must be exercised in making the walls of the cup of uniform thickness if a satisfactory product is to be obtained.

Annealing and Redrawing Operations. — After the cups are made, it is the general practice to anneal, wash, and dry them as described in Chapter III. Then they are ready for the first redrawing, or second drawing operation. The temperature to which the cup is heated for annealing varies from 1200 to 1220 degrees F. The manner of handling the cups after they have been annealed, washed, and dried, is to carry them in trucks, which are lifted from the floor of the annealing room to a track located above the drawing presses. These trucks are provided with false bottoms and are run along the track until they are directly over the hopper which feeds the cups to the punch press. The false bottom is then removed, allowing the cups to drop from the chute into the hopper A of the drawing press, Fig. 6, from which they are removed by a feeding device consisting of a wheel in which pins B are set at an angle of about 45 degrees with its horizontal axis. These pins are pointed, enabling the case to be located on them, mouth first. The pins are rotated inside the hopper so that they catch the cups and deposit them in close-wound spring tubes C. These tubes pass from the hopper down to the feeding slides of the drawing press, and as the cases drop out of the tubes they are caught by fingers held on the slides and carried over into line with the dies and punches. When the slides have advanced to their extreme forward positions, the punches descend and force the cups through the drawing dies, depositing them in a box located under the press. The slides are operated from the crankshaft through bevel gears and a connecting-rod that transmits power down to a horizontal shaft carrying a series of four cams. These cams are in contact with rollers held in the feeding slides and thus transmit the desired movement to them. The rolls are held in contact with the cams by coil springs. The machine shown in Fig. 6 operates at 100 revolutions per minute, and as four cups are drawn per stroke, it is evident that this machine has a productive capacity of 24,000 cups per hour. The drawing

Fig. 7. Duplex Drawing Press performing Fourth Redrawing Operation and turning out 10,800 Cups per Hour

press shown in this illustration is used for performing the first, second, and third redrawing operations, shown with the die and punch used at *A*, *B*, and *C* in Fig. 1.

Fourth and Fifth Redrawing Operations. — The fourth and fifth redrawing operations on the case are handled in machines of a type similar to that shown in Fig. 7, which are provided with only two punches and dies instead of four, as was the case with the machine shown in Fig. 6. The feeding of the cases to the slide that carries them to the dies is practically identical with that shown in Fig. 6, but the slide is operated in a different manner. In this particular machine the slides A, which serve as a means for carrying the cups from the feeding tubes B over into line with the drawing dies, are actuated in their movement by means of a bellcrank lever receiving power from a cam held on the crankshaft C of the press. While the cases are fed to the punch with the mouth up, it sometimes happens that one will pass down the feeding tubes to the slide the wrong way, that is with the bottom up. Now should such a case be allowed to pass over into line with the die, it would mean that the punch would be broken and the die either broken or damaged to such an extent that it would be unfitted for use. It is not uncommon also to have cases pass down to the slide that are dented or otherwise defective which would prevent them from feeding into the die properly. Should such a case pass down the feeding tubes and stick in one of the slides, it would mean that the punch would come down on the slide and break, not only putting the machine out of commission for a time, but perhaps causing serious injury to the attendant as well.

In order to provide against such accidents, an ingenious tripping device is applied to this machine. This device, while comparatively simple in construction, is positive in its action, and has been the means of saving a great deal of money in the cost of dies and tools. It also enables one attendant to run four instead of two machines. Essentially, this device consists of a projecting stud held in the crankshaft of the press, and which when the feeding slide is operating normally passes through a slot cut in a lever that is connected with the bellcrank lever operating the slide. Now if, for any reason, the slide should be prevented from making a complete forward or backward stroke, this projecting pin would not pass through the slot in the lever mentioned, but would force the lever out, knocking out the lever D, which transmits a movement through the links *E* and *F* and bellcrank *G* down to the tripping lever *H*. This knocks the clutch operating lever *I* off the catch—throwing in the clutch and stopping the operation of the press. It can therefore be seen that this tripping device is of simple construction, but is effective, owing to the fact that when the slide

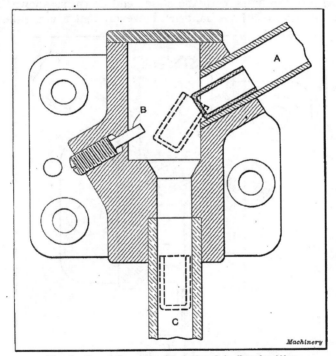

Fig. 8. Cup delivered to Feed-pipe C in Regular Way

does not complete its movement the clutch is thrown in before the ram of the press reaches the top of the stroke so that the machine is stopped before it has a chance to complete another stroke.

Feed Chute for Drawing Press. — Another interesting device for preventing the breakage of punches and dies is shown in Figs. 8 and 9. The press on which this device is used is employed in performing a drawing operation on the cups shown in the illustration. These cups enter the chute A from a hopper and pass down at high velocity; when the bottom of the cup strikes the pin B the cup rebounds and drops into the feed-pipe C, which leads to the feed fingers of the press. It is necessary to have the cups feed into the press with the closed end down, as shown in Fig. 8. It sometimes happens, however, that a cup enters

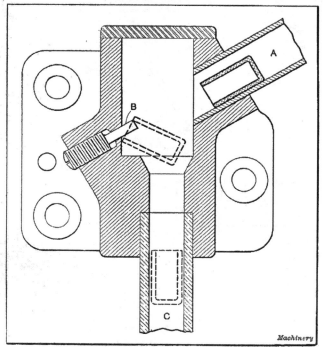

Fig. 9. Inverted Cup In Act of Turning Over Before Entering Feed-pipe

the chute A in an inverted position, as shown in Fig. 9. If the cup reached the die in this position, it would result in breaking the drawing punch. Fig. 9 shows how this is avoided. The cup enters the chute A and slides down until it catches on the pin B. Thus it cannot rebound, but swings

on the pin *B* and drops into the feed-pipe *C* with the closed
end down in the manner illustrated. This device is carried
by a bracket which is attached to a stationary part of the
press.

Final Redrawing Operation. — The fifth redrawing operation is accomplished in a press similarly equipped to that
shown in Fig. 7. These presses operate at 90 revolutions
per minute, and turn out 10,800 cups per hour. Several
annealing operations take place between the time when the
cup leaves the first redrawing operation and the time when
it is ready for trimming, but, as these have been described
in the preceding chapters, no reference need be made to

Fig. 10.   A Battery of Automatic Trimming Machines at work
on 0.30-caliber Cartridge Cases

them here. Before the fourth redrawing operation is accomplished, the cases are taken to a heading machine of
the horizontal type where they are "bumped." This operation is accomplished in order that in the successive redrawing operations the head of the case will not be reduced too
much in thickness. The 0.30-caliber cartridge case has
what is known as a solid head; that is, the top portion of
the case that contains the primer is not intended to form a
pocket for the primer, the pocket itself being simply a hole
forced into the head. This type of cartridge has been found
ncessary for use with smokeless powders. · The former

method used in making 0.30-caliber cartridge cases was to form the pocket by forcing in the head which was very little thicker than the sides of the case near the head. This construction, however, was found to be too weak for smokeless powders, as the head would blow off. The "bumping" is a very simple operation and is somewhat similar to heading except that the punch is perfectly flat and simply gives the case a blow, upsetting it slightly and flattening it so that in the two final redrawing operations—fourth and fifth —the metal at the head is not stretched to any appreciable extent. The cartridge case is now ready for trimming.

Fig. 11. Closer View of One of the Automatic Trimming Machines shown in Fig. 10

Trimming a Cartridge Case to the Exact Length. — As mentioned, it is practically impossible to draw a case that will not have an irregular top edge and that will not become distorted or cracked to some extent at the edge. This makes it necessary to draw the case much longer than actually required, and to trim off the surplus material. The removal of this excess amount of stock is accomplished in machines that are operated automatically. A battery of these trimming machines at work on 0.30-caliber cartridge cases is shown in Fig. 10, while Fig. 11 shows a closer view of one of the machines and gives a clear idea of its working mechanism. As will be seen upon reference to Fig. 10, these trimming machines are arranged in such a manner that the various hoppers can be filled from an overhead conveying system. This arrangement consists of a track similar to that used in the drawing press department previously referred to, and enables one man to attend to an entire line of presses. The track accommodates a truck in which the cases are carted along the line and from which they are ejected through a false bottom, dropping into the hoppers located over the machines. The feeding of the cases down to the trimmer is accomplished by the same type of hopper as previously described, but the subsequent handling is somewhat different. As the case descends from the hopper, it passes through a locating cage A from which it is carried forward by a plunger B and is located on the cutting-off punch C. Here it is held by friction while a circular trimming tool D advances and trims off the surplus stock. The case and trimming are then ejected from the cutting-off punch by a sleeve E operated from the left-hand end of the machine, the case being deposited in one box and the trimming in another; two separate channels are provided as shown clearly in Fig. 11. Following the trimming operation, the case is headed and the mouth annealed and reduced. These operations have been completely described in Chapter IV.

CHAPTER VI

MAKING DIES FOR CARTRIDGE MANUFACTURE

WHILE the making of combination blanking, cupping, and drawing dies for cartridge manufacture does not differ materially from ordinary die-making, there are a few points in connection with this work that it might be well to explain. The die blank is made from a special grade of Firth-Sterling steel containing from 1.11 to 1.30 per cent carbon. This is an extremely high-carbon steel, but has been found satisfactory for this class of die, owing to the great wear

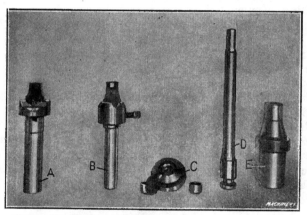

Fig. 1. Combination Blanking, Cupping, and Drawing Punch and Die, and the Tools used for Making the Die

to which it is subjected when in use. The first step, of course, in making one of these dies is to cut off the blank from a bar of stock, and then by means of drills, reamers, etc., to shape the hole in the die to the correct form. Fig. 1 shows one of these dies at C, together with a blank and a cup made from it, while D is the combination cupping and drawing punch and E the blanking punch. Fig. 2 shows a close view of the chuck used for holding one of these dies. The manner in which the die is held while being drilled, counterbored, and reamed is identical with that used when it is set up in the press, so that in this way the conditions in both cases are as nearly alike as possible. The chuck consists of a female center A in which a recess is provided that fits the external body of die B. The outer end of this center is turned and threaded to fit a cap C that is machined to correspond with the other end or smallest diameter of the die B. This cap holds the die rigidly in position while it is centered with the female member A. In addition to a drill, and, of course, a boring tool to true up the hole, two tools are used for finishing the hole in this die. The first

Fig. 2. Chuck used for Holding Combination Blanking, Cupping, and Drawing Die Blank when Machining the Hole

or roughing tool B, Fig. 1, is of the flat type, having one cutting edge, while the finishing tool A is of somewhat similar shape but has considerably more circumference so that a rounder hole will be produced. The first tool is used merely for roughing out purposes and for bringing the hole to its approximate shape.

Making Redrawing Die Blanks. — The blanks for redrawing dies are made in Cleveland automatic screw machines and are ready for the final reaming operation when they drop from the machine. This method is commendable in that it reduces the cost of the die blanks to a minimum.

Die *G*, illustrated in Fig. 3, is a second redrawing operation die for a 0.30-caliber cartridge case, and is 1 23/32 inch in diameter by ½ inch thick. In order to have the hole true with the external diameter, great care is taken in spotting the work and then removing the hole entirely from the next blank, using a fairly wide cut-off tool. The order of operations accomplished in the proper sequence is as follows: First, feed stock to stop *A*, Fig. 3; second, turn external diameter with box-tool *B* held in turret, and spot with a drill *C* retained in the same holder; third, drill with drill *D*; fourth, ream straight portion of hole with reamer *E*; fifth,

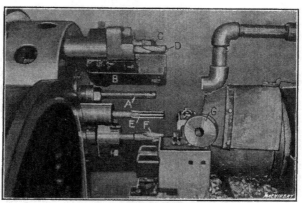

Fig. 3. Making Redrawing Die Blanks in a Cleveland
Automatic Screw Machine

bell-mouth with reamer *F*, and face with a tool held on the rear cross-slide; sixth, cut-off with a tool held on the front cross-slide.

The drawing dies, after being rough-formed in the manner illustrated, are taken to the tool-room where they are reamed out to the exact diameter and bell-mouthed to the correct shape, after which they are ready for hardening. For hardening, the dies are heated in a muffle furnace to a temperature varying from 1400 to 1450 degrees F., and are then "spouted" as illustrated in Fig. 4. The spouting of the die consists in directing a stream of water through the hole in order to harden it and at the same time leave the external diameter practically soft. The reason for this is that the die, after hardening, is not drawn, and if the entire blank were hardened it would break very easily. Having the external diameter soft increases the strength to a considerable extent, and the dies wear much longer and do not break as easily. When spouting, the die is held in a cage formed in the base of the bracket A and then the funnel B (which is similar in shape to an ordinary oil funnel except that the lower tapered tube is left off) is placed over it, the

Fig. 4. "Spouting" a Redrawing Die—Hardening the Hole and
Leaving the Exterior Comparatively Soft

water being directed through this funnel and thence to the hole in the die. The funnel is provided with a handle to enable the operator to place it quickly in position over the die after the latter has reached the proper temperature, and has been placed in the fixture. The operator removes the die with a pair of tongs, holding the tongs in one hand and the funnel in the other.

Great care must be taken in heating this grade of steel because of its high carbon content. A variation of 10 degrees F. one way or the other from the temperature found most satisfactory is claimed to make the die defective. It has also been found necessary to heat the dies for various operations to different temperatures; that is to say, the die that would be used for a first redrawing operation would be heated to a different temperature from one that would be used for the fifth redrawing or final drawing operation. The reason for this is that the pressure exerted on the die, by forcing the cup through it with the punch, on the first drawing operation, is much greater than that for the final drawing operation, and hence the die cannot be as hard and

Fig. 5. Lapping a Redrawing Die

must have a more porous grain to withstand the additional pressure.

Lapping Redrawing Dies. — After hardening, redrawing dies are lapped in the manner illustrated in Fig. 5. The die is held in a holder, resembling somewhat in shape that used in the drawing press, which is held in the chuck of the speed lathe. For lapping, a lead lap is used. This is held on a steel plug provided with a handle driven through it at right angles to the lapping portion, so that the operator can grip it with both hands and thus hold the plug in the proper position in relation to the hole in the die. The speed lathe in which the die is held is operated at from 2000 to 2400 revolutions per minute, and the spindle is always kept a snug fit—also there must be no end play. A mixture of lard oil and No. 10 emery is used for lapping, this having been found best for both the roughing and finishing operations. The mouth or bell-mouthed portion of the die is lapped with emery cloth of the same grade—No. 10. The lapping of a drawing die is an operation that must be carefully handled. Not only must the hole be of the correct size, but the radius of the bell-mouthed portion must be exact. The correct lapping of the die is more a matter of experience than anything else, and it is practically impossible to give any definite information on the subject. One point, however, that should never be ignored is the fact that the lap should always be presented in a line parallel to the axis of the die. If it is tilted over the least bit to one side or the other, a hole will be produced that will not only be out of round, but that will not be straight; that is, if the die were placed on the arbor it would be found to run untrue because the hole would not be exactly in the center of the blank on both sides.

After the dies are lapped to the correct size and shape, they are ready for use in the press and are then turned over to the drawing press department. Drawing dies for all redrawing operations up to the final operation are used until they have worn approximately 0 0017 inch large. When they have become worn to a size this much greater than the actual diameter of the cup required, they are taken out of the press and annealed. The dies are then reamed out to the next size larger—that is for the redrawing operation preceding that for which they were originally made—and used over again. This is repeated until the dies have been used five times. Redrawing dies made from Firth-Sterling steel of the carbon content mentioned are good for making 40,000 cups before they have become worn too large. The peculiar point about this steel is that it does not warp out of shape in hardening and also does not produce any scratches on the work. It is of extremely fine grain, hardens well and produces a case free from scratches and other imperfections. The only thing that makes it unfit for use is when it becomes worn too large. Otherwise the condition of the hole in the die is as good at the completion of 40,000 cups as it was when first used.

Making Redrawing Punches.— Redrawing punches are also made from Firth-Sterling steel, but the carbon content for the punch is much lower than that used for the die, and should never exceed over 0.60 per cent. There are a few points that require careful consideration in making drawing punches. In the first place, the stock should be centered as true as it is possible to make it. There is a good reason why this operation should be carefully done. If the piece of stock from which the punch is to be made has not been centered true, the finished drawing punch when hardening will be bent out of shape. The reason for this is that when a bar that has been incorrectly centered, so that it runs eccentric, is turned, more stock is removed from one side than the other, and the turning, instead of being parallel with the grain, cuts across it; hence, the ability of the steel to resist deflection in hardening is not as great as it would be had the support not been removed from one side. The explanation given for this is that in rolling bar stock the fiber or grain of the metal is drawn out in practically a straight line and when this condition does not exist in the finished article distortion takes place, because, in cooling, the fibers of the stock revert to their original positions parallel with the axis of the bar.

Another point which is of considerable importance is that never less than 1/32 inch of material should be removed from the bars if the finished piece made from it is to be hardened. There is a certain decarbonizing portion surrounding a bar of stock that prevents the steel from hardening properly, and this decarbonizing portion should always be removed from those parts of the punch that must be hardened; if not, soft spots will be experienced. The drawing punch should be heated in a muffle furnace very slowly until it has obtained the correct temperature, and while being heated it should be constantly rotated to prevent warping. The temperature to which drawing punches are heated varies from 1400 to 1425 degrees F. They are quenched in a bath consisting of 15 parts of water and 1 part of common potash, and are dipped in a vertical position as illustrated in Fig. 6.

Cluster Double-action Punches and Dies.— Combination blanking punches and dies are made by many manufacturers in cluster forms so that more than one cup is made at each stroke of the press. While these dies have many advantages over the single type construction, they require

Fig 6. Hardening a Redrawing Punch—Note Method of Dipping

great care in the laying out and making. For instance, the holes in the die-blocks must be reamed approximately 0.005 inch larger than the collets and are transferred to the die-block through bushings fitted into the punch-block when the two members are clamped together. If the holes for the die-holders are not laid out accurately, the tie or web between the blanked holes will be cut out as the metal feeds through, any error in spacing, of course, being multiplied as the work progresses. However, if the die-holder is properly laid out, it is possible to cut very close on scrap. On thin metal, in actual running, it has been found that the scrap averaged 18 per cent, while the estimated scrap was 18¼ per cent. The difference between the actual and estimated scrap was due to the slight amount that the feed-rolls squeezed the scrap back, making the webs slightly less than that estimated.

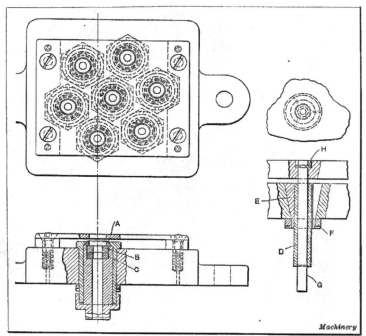

Fig. 7. Cluster Double-action Punches and Dies

As a general rule, the punch- and die-block for this type of die is made from steel, and while this is not necessary as a matter of strength, it does away with the chances of striking a blow-hole in cast iron. The collets, sleeves, and nuts, forming the members for holding the dies, are also made from tool steel, hardened and ground. The blanking dies *A*, Fig. 7, are made from high-carbon steel heated to the recalescent point, cooled in brine, at 62 degrees F., then placed in an oil bath at 225 degrees F., after which they are ground. Each blanking die is good for approximately 1,000,000 blanks before grinding on the top face, which can be done, as a rule, about ten times, if necessary on account of wear and not due to chipping. It can be seen by referring to Fig. 7, from the way in which the blanking die is held, that if the die is cracked from any cause, after it is ground to size, it still can be used without detriment to the blank. The sleeve holding it is tapered, and, in this way, as it is

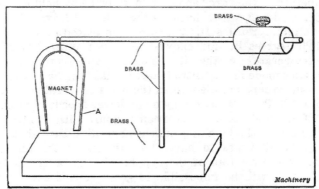

Fig. 8. Magnetic Device for Determining Recalescent Point of Dies when Heating for Hardening

drawn up by the nut under the bolster, the die is securely held in place, and, if cracked, would be closed up.

The cupping dies *B* and *C* are made by one manufacturer from "Intra" steel, which is used because of the extremely hard surface obtained when it is heated to the proper temperature and "spouted." It is also extremely tough when heated in this way, and the combination of hardness and toughness is ideal for cupping dies. The small size of these dies makes it necessary to use a steel of great tensile strength, as well as one that will harden well, and "Intra" steel fills these requirements.

The cupping dies are heated to the rescalescent point, which is found by using the testing fixture shown in Fig.

8. The die to be tested is placed under the magnet *A* from time to time, as the heating progresses, and when the point is reached where the steel has no attraction for the magnet, the heat is noted on a temperature pyrometer, and all dies made from this grade of steel are heated to that point. If a temperature pyrometer alone is used, without the magnet, readings are taken at stated intervals of approximately one minute apart, and then a chart is laid out by taking the readings as obtained from the pyrometer. This chart, if properly laid out, will show at the recalescent point a horizontal line. When the cupping dies have reached the proper temperature, they are "spouted" with brine by holding them under a funnel, which hardens the hole and leaves the outside circumference soft. They are then placed in oil, heated to 225 degrees F., and allowed to remain until they reach the temperature of the oil. One noticeable feature of cupping dies made from "Intra" steel is that they have a longer life—more cups per die—when the press is operated at from 125 to 135 R. P. M. than they do on lower speeds of from 90 to 100 R. P. M.; also the cups come square with less trouble.

In making the blanking punches and dies, care should be taken at all times to have the bottoms of the blanking punches and the tops of the cupping dies parallel. If they are not, it will be impossible to get a square cup, as the punch will bear harder on one side than on the other. In reaming out the hole in the cupping die, care should be taken to see that it is perfectly round and square with the top and bottom faces. A good method of determining if such a requirement has been obtained is to wring the plug gage into the die; this leaves a ring around the hole in the die parallel with the top face when the hole has been properly machined. That is to say, after swabbing out the die and cleaning the gage, a slight oil film is left on both parts; then when the gage is wrung into the die the liquid constituents of the oil are squeezed out, leaving a small carbon deposit. This should show heaviest just at the point where the drawing portion of the die starts to round off into the straight or sizing portion. On the lower or stripping die, it is necessary to have the line show heaviest at the bottom, as a slight bell-mouth on the bottom will make the cup strip hard.   If this line left by wringing the gage into the die is not parallel with the top face, the cup produced will not be perfectly straight, but will have an irregular top edge. This undesirable condition generally results from a chip clinging to the cutting edge of the reamer for a portion of a revolution when reaming out the die.   It is also often caused by poor lapping, but more generally by reaming, as previously mentioned.   If more attention were paid to this point there would be less trimming with its attendant scrap on redraw work, as it is practically impossible to get a floating die to draw square unless it has a perfect "ring" bearing.

The blanking punches D, one of which is shown to the right in Fig. 7, are held in the punch-block by the split sleeve E, which, in turn, is drawn down by the nut F.   This makes it possible for the punch D to slip up if more than one blank gets under it, which sometimes happens when a poor piece of metal gets into the press, and the drawing punch G punches out the bottom of the cup instead of drawing it through the dies.   Both the blanking and drawing punches are made from high-carbon steel, heated slowly to the recalescent point in a muffle furnace and rolled around from time to time to get an even heat.   They are then plunged in water that has previously been heated to 80 degrees F. This is done slowly in a tank which is so deep that the punch will turn black before being withdrawn; that is, the tank should have sufficient depth so that the punch, when dipped straight, will not touch the bottom before it has been cooled sufficiently to harden; it is also withdrawn straight from the tank.   After hardening, the punches are then dipped in oil, heated to 225 degrees F., and allowed to remain in the tank for from ten to fifteen minutes.   Afterwards they are ready for grinding.

The drawing punch is made with a taper on the end of from 0.004 to 0.005 inch to the inch of length, and is ground to this taper for approximately one inch.   When the tapered ends of the punches have worn below the required size, they are cut off on the end and re-ground, the slide of the press being lowered to compensate for the reduction in length   The upper ends of the drawing punches are held by slotted washers *H* which are fitted into counterbored holes in the punch-block. The cluster double-action punches and dies shown in Fig 7 are used in double-action crank and double-action cam presses, but for work using metal thinner than 1/32 inch the cam press is preferable on account of the dwell which it is possible to obtain at the lower end of the stroke while the drawing punches are doing their work.   With a strong, well-designed and well-built press, no trouble should be experienced in cutting with a minimum amount of scrap, when 'the tools and feeding device have once been properly adjusted.

CHAPTER VII

MAKING SPITZER BULLETS

THE methods used in the manufacture of spitzer bullets for 0.30-caliber cartridges are of an unusually interesting nature. The point of the bullet must be absolutely concentric with the body if good results are to be obtained. If the point is slightly eccentric, the bullet is erratic in its flight and cannot be depended upon to shoot accurately. It is, therefore, not only necessary to exercise the greatest care in the final operations on the bullet, but the same exactness and careful attention must be followed from the initial operation through to the finished product. The cup, in fact, has to be started just right. The walls must be of equal thickness and drawn straight, as any difference in the length of the cup means, as a rule, a variation in the thickness of the wall. This variation in the thickness of the wall will be apparent in the finished jacket and is somewhat exaggerated in each successive operation, so that the care that must be exercised in drawing makes this work rather unusual in its character.

Making Cups for Spitzer Bullets. — The cups from which spitzer bullet jackets are made are produced from stock composed of 85 per cent of copper and 15 per cent of nickel. This material is purchased in sheet form and comes in rolls about 12 or 14 inches in diameter. In operation, the roll is held on a stand, as was described in connection with the making of the brass cartridge case in Chapter V, and is fed into a double-action press. When making the cups for the spitzer bullet, five combination punches and dies, as shown in Fig. 1, are used. These are of the same shape and type as those shown in Fig. 1, Chapter VI. The press is operated at a speed of ninety-four revolutions per minute, giving a production of 470 cups in this time. In addition to blanking out and cupping, this machine also performs the first drawing operation on the nickel cup. Before any further operations are performed, the cups are annealed, washed, and dried and are then ready for the redrawing operation.

Redrawing Operations on Nickel Cases. — The redrawing operations on the nickel cases for spitzer bullets are accomplished in a drawing press of the type shown in Fig. 2. This machine carries two punches and two dies as illustrated. The punches are carried in a head operated by a crankshaft, and the feeding mechanism is actuated by

Fig. 1. Blanking, Cupping, and Drawing the Nickel Case for Spitzer Bullets at the Rate of 470 per Minute, using Five Combination Punches and Dies

a cam, receiving power from the crankshaft through a pair of bevel gears and the vertical shaft shown. The cups are placed in a hopper located at the top of the machine in which a wheel carrying pins rotates. These pins pick up the cups and carry them to a receiving chute, from which they pass down through brass tubes made from close-wound brass springs, which carry the cups down to the feeding mechanism of the machine. The feeding mechanism consists of a slide in which fingers are held that grip the cups and carry them forward to the punches, which as the ram descends force the cups out of the fingers and through the redrawing dies. There are three redrawing operations on the nickel case as shown at *A, B,* and *C* in Fig. 3; this illustration also shows the three redrawing punches and dies used. The redrawing operations are accomplished in a press of the type shown in Fig. 2, which operates at 112 strokes per minute and turns out 224 cups in this time.

Making Lead Fillings or Slugs. — The lead fillings or slugs for spitzer bullets are made from wire which is produced in the hydraulic press shown in Fig. 4. This lead wire is made from large billets of a composition metal consisting of 30 parts of lead and 1 part of tin. A group of these billets just as they come from the mold may be

Fig. 2. Type of Drawing Press used for Redrawing Nickel Jackets for Spitzer Bullets using Two Punches and Dies

seen in the foreground of the illustration. These are cast in an iron mold and are 4¾ inches in diameter by 13¾ inches long. The machine in which the billets are placed is operated on the hydraulic principle, water being used as a medium for obtaining the desired pressure, and is capable of exerting a pressure of 1000 tons per square inch. In operating this machine, the lead billet is placed on the stand

A, and, as the plunger of the machine ascends after forcing out the metal, this stand also rises until the billet comes almost level with the chute *B*. The operator then disaligns the plunger, as will be described later, dumps the lead billet off the chute, dropping it down into the die of the machine, aligns the plunger, starts up the machine, and as the plunger again descends, it forces the metal out through sizing dies. The metal, in passing out from the large die or billet container of the machine, passes through sizing dies which form the wire to the correct diameter, in this case, 0.380 inch. When the machine first starts to force out the lead

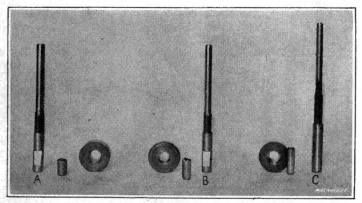

Fig. 3.   Sequence of Three Drawing Operations on Nickel Case for
Spitzer Bullets; also Dies and Punches used

wire, the gage indicates a pressure of 600 tons, which gradually drops down to 550 tons as the metal begins to heat up and becomes softer.

As the lead wire is squirted out of the dies, it passes over a roll *C* and thence through a box *D* filled with constantly running water. This serves to cool the wire. From the box of water the wire passes over another roll and up through an adjustable mechanism which is used for winding the wire on the spool *E*. The mechanism for locating the wire on the spool is controlled by a double-pitch worm cut right- and left-hand which guides the arm of the feeding or winding attachment. As the wire in being wound

Fig. 4. Hydraulic Press for Making Lead Wire which is used as Filling for Spitzer Bullets

up travels over to one side of the spool, the direction of
the feeding or winding arm is changed by a dog which
throws the nut into the groove in the worm running in the
opposite direction. The worm is driven by gears from the
pulley *F*, and the spool is driven from the same pulley
through a friction clutch.

To insert the lead billet in the central die of the machine,
it is necessary to throw the punch out of line with the die
so that the billet can enter. This is done in order to decrease the amount of punch travel that would otherwise
be necessary, and allows the machine to be operated much
more quickly. The mechanism for disaligning the punch is
of very simple construction, but is fool-proof. The plunger
G is thrown out of line by operating the handle *H* and is
brought back into line by operating the handle *I*. By look-
ing carefully at this illustration, it will be seen that handle
I is connected with the mechanism operating the valve of
the hydraulic pump, and hence it is impossible to start the
plunger on its downward stroke until it is in perfect alignment with the die. This attachment prevents any accident
which might happen should the operator fail to align the
punch accurately with the die and then start the machine.

Swaging Lead Slugs. — The lead wire produced in the
hydraulic machine shown in Fig. 4, after it has been wound
on the spool, is taken to the swaging machine shown in
Fig. 5, the spool being placed on a stand as illustrated at *A*.
The wire passes from this roll down over a number of
other small guiding rolls and is forced up by a feeding device located under the machine. This feeding device lifts
the wire at each stroke an amount sufficient to produce two
slugs. The machine is provided with two dials *B* and *C*,
one over the other. The dies in these dials are not used for
forming the slugs, of course, but merely act as feeding or,
in other words, carrying members. One die in each of the
dials *B* and *C* is successively in perfect alignment with the
corresponding one in the other dial when the machine is
indexing properly. The top dial, which is geared to the
lower one, is one-half the size and carries one-half as many
dies as the lower one. The lower dial indexes two positions

for one index of the upper dial, the latter serving to feed each second successive hole in the lower dial. The lower dial, which makes two indexes for every one made by the upper dial, is used for carrying the slugs to the various punches and dies that are used for forming them to the proper shape. There are two sets of punches and dies; the first two punches *D* are used for forming the slugs, the

Fig. 5.   Swaging Press used for Making Lead Fillings for
Spitzer Bullets

second punches *E* force the formed slug out of the dies, while the two punches *F* are used as a means of safety. The punch *G* is attached to the knock-off and stops the machine should the wire be feeding short. The last punch *H* knocks out any scrap that may remain in the top dial, and keeps the dies clear. The dies held in the lower dial are used in conjunction with bushings in the machine for cutting off the wire as it is fed up; then the amount of wire fed at each stroke is again subdivided by the dies held in the top and lower dials that work against each other as they rotate. This machine works entirely automatically and, when properly set up, produces slugs at a rapid rate.

Finish-forming Spitzer Bullets. — The most difficult operation in the production of spitzer bullets is the finish-forming, which consists in pointing the nickel case and making the point absolutely concentric with the body. The type of machine used for this purpose is shown in Fig. 6, where the sequence of operations on the bullet is shown by

Fig. 6. Machine used for Forming the Nickel Cases for Spitzer Bullets and Assembling the Lead Fillings in them

the samples on the dial. This machine is of the ratchet dial feed type. The nickel cases are located in bushings held in the dial A by the operator and are carried around by this medium. The slugs which are held in a hopper at the back of the machine are fed down to the second dial B, through a close-wound spring tube. Dial B is indexed in a similar manner and in unison with dial A, and the slugs which are carried in it are forced out and located in the nickel case in the lower dial by means of punch C held in the ram of the machine.

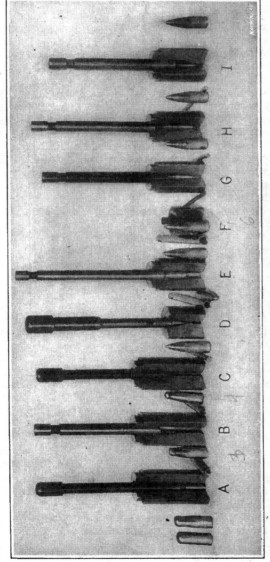

Fig. 7. Illustration showing the Type of Dies and Tools used in the Machine shown in Fig. 6 for making Spitzer Bullets

Fig. 7 shows the sequence of operations followed in the production of the spitzer bullet, and also the dies, punches, and knockout pins used. The dies in this case have been sectioned so that their shape can be clearly seen. Figs. 8 and 9 show diagrammatically the operations performed on the spitzer bullet. Here the relations of the forming punches and knockout pins to the dies are clearly illustrated. By referring to these illustrations, the sequence of operations on the spitzer bullet will be clearly understood. At A the case is partly reduced in diameter at the point. At B the point of the case is still further reduced and the rounding of the body is commenced. At C this operation is carried still further, while at D the case is completely formed. In all the operations up to this point, the case is formed to the shape shown by the punch and die working in conjunction with each other; that is to say, the punch is the same shape as the die, but is smaller in diameter an amount equal to twice the thickness of the walls of the case. At D the punch is relieved and only fits the case at the lower or curved portion; however, the punch follows the case right to the point and is used in connection with the die for forming the point concentric with the body. At E the lead slug is inserted in the nickel case, and it will be noted that the punch cannot pass down into the case, but merely rests on top of the lead slug. At F the operation accomplished is known as "coning." This is one of the most important operations, as it is intended to make the point of the bullet perfectly concentric with the body. The die a which is used for this purpose only forms the extreme point of the bullet, and does not bear at all on the curved part of the body. These pointing dies are only allowed 0.002 inch float in the holder and must be perfectly in line with the punches. The punches and dies shown from G to I are used for forming the base of the bullet. These operations consist in turning over the top edge of the nickel case and flattening it down on the lead slug—J and K show the condition of the nickel case before and after pointing, while L shows the lead slug and M, the finished bullet.

The proper setting up of the machine shown in Fig. 6 is an operation that requires some skill. The punches which are held in the ram of the machine must be in absolute alignment with the dies. The pointing dies are only allowed 0.002 inch float in the holders in which they are retained, and are set as exact as it is possible to get them. After the machine has been set up, a number of trial bullets are run through; these are then taken out and tested in a

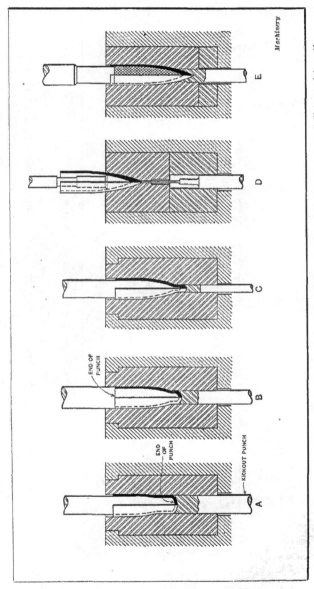

Fig. 8. Diagrammatical View showing the Sequence of Operations in the Production of Spitzer Bullets, and how the Various Tools act in Relation to each other

Machinery

98

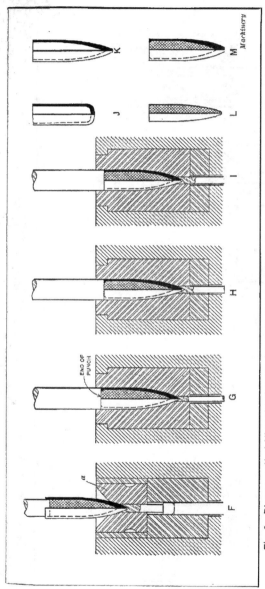

Fig. 9. Diagrammatical View showing the Sequence of Operations in the Production of Spitzer Bullets, and how the Various Tools act in Relation to each other

special type of high-speed lathe in which the bullet is held true. When put under this test the point of the bullet must run perfectly concentric with the body before the machine can be started on a "run."

This test of the bullet is also made from time to time so that any variation in the setting of the dies will be detected before many scrap bullets have been run through. The making of the dies for bul-

lets of the spitzer type is considered by cartridge makers to
be one of the most difficult propositions in cartridge making.
It requires the most excellent tool work and also special
care in using the tools, setting them up in the machine, and

Fig. 10. Machine used for Canneluring Spitzer Bullets; also
for Sizing them

keeping the machine in absolute alignment and perfect running order.

Canneluring and Sizing Spitzer Bullets. — In order to
provide a means for holding the spitzer bullet in its proper

position in the cartridge case, the bullet is provided with a knurled groove around its periphery near the base. This operation is known as canneluring, and is accomplished in the machine shown in Fig. 10. The bullets are put into the hopper *A* from which they are conveyed by means of a tapered conveyor screw *B*. Conveyor *B* is rotated by means of a belt *C* running over the pulleys illustrated. The conveyor carries the bullets up to the entrance of the chute *D* which is a close-wound spring, connected with the small

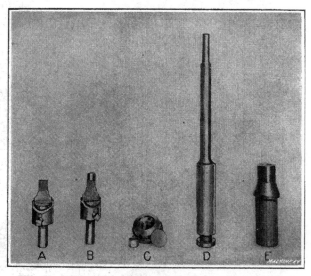

Fig. 11. Combination Blanking Punches and Dies used for making the Nickel Jackets and Tools used for making the Die

hopper forming an auxiliary to the six-hole turret *E* located above the canneluring dial *F*. Turret *E* in which the bullets are dropped rotates and carries the bullets around so that they can be inserted in the dial *F* by means of punches. Dial *F* is also rotated and carries the bullets past a knurling segment, which produces the small groove in the circumference. As the bullets are cannelured they are forced out of the top dial *F*, pass down through a tube and are located in the lower dial *G*. This lower dial carries the sizing dies, which resize the bullets, this operation being necessary owing to the slight distortion produced during the canneluring operation. The punches for forcing the bullets through the dies held in dial G are carried in the slide H which is operated by crankshaft I. The dial G is rotated by a ratchet mechanism receiving power from a fulcrumed lever J which carries a roll working in and receiving motion from a groove in cam K.

Making Dies for Spitzer Bullets. — The type of tools used for making combination blanking, cupping, and drawing dies for spitzer bullets is shown in Fig. 11. These tools are of a type similar to those illustrated in a former chap-

Fig. 12. Making the Pointing Die for a Spitzer Bullet in a Hand Screw Machine

ter. Here C is the combination blanking, cupping, and drawing die, E the blanking punch, and D the combination cupping and drawing punch, while A and B are the roughing and finishing reamers, respectively, for the die. The information given in regard to the dies for making the brass case also applies to these dies.

Tools used for Making Pointing Dies. — Fig. 12 shows how the pointing dies for spitzer bullets are made. Sanderson hammered steel is used for making these dies because of its "shock resistance" and also the great hardness that can be obtained without danger of cracking or breaking. The operations in the making of this die, which is

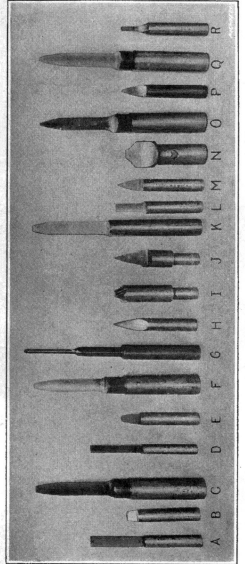

Fig. 13. Type of Tools used for Making the Dies for Producing Spitzer Bullets, these Dies being used in the Machine shown in Fig. 6

shown at *D*, Figs. 7 and 8, are as follows: Drill large portion of hole; drill small portion of hole with a No. 68 drill, and insert roughing and finishing pointing reamers. The making of this die represents an excellent example of tool work, as the holes must be in absolute alignment with each other, and the entire curved portion of the hole concentric throughout.

Fig. 13 shows all the reamers that are used for making the pointing dies shown in Figs. 7

and 8. The names of these are as follows: A is the reamer for the first die; B is the first and second operation stem reamer; C is the reamer for the first and second die shown at A and B in Fig. 7; D is the second operation die reamer; E is the second operation stem reamer; F is the third operation die reamer; G is the reamer for the small hole in the third operation die; H is the first pointing die reamer; I is the countersink for the mouth of the pointing die; J is the pointing die finishing reamer; K is the "slug die" reamer; L is the slug die stem reamer; M is the coning die stem reamer; N is the coning die reamer; O is the hump die reamer; P is the hump and base stem reamer; Q is the base die reamer; and R is the base die stem reamer. It will be noticed by referring to this illustration that practically all the reamers are of the "spoon" variety. This type presents only one cutting edge to the work, but is easily made and can be gaged with less difficulty than a fluted reamer. It is also readily sharpened, being stoned only on the top flat face.

CHAPTER VIII

LOADING AND "CLIPPING" CARTRIDGES

In Chapter V, the development of 0.30-caliber cartridge cases from the sheet to the finished case was described, the methods recorded being those used by the Frankford arsenal. The operations referred to were the blanking, cupping, drawing, and trimming of the cases. The heading operation was described in Chapter IV, in connection with methods used by the Dominion Cartridge Co.; hence it will not be necessary to deal with this subject here, as the methods followed in the Frankford arsenal are practically the same as those used by the other concern. The method of priming 0.30-caliber cartridge cases in 'the Frankford arsenal, however, is different from that used by the Dominion Cartridge Co., and a description of the priming of the cartridge, that is, the placing of the detonating cap in the head of the cartridge case, will, therefore, be given here.

Priming. — The head of the cartridge case, when formed in the heading machine, is provided with a pocket, which is made by a teat on the end of the heading bunter. This pocket is approximately of the same size—slightly smaller —than the primer. The inserting of the primer in this pocket is called priming. It is done in the machine shown in Fig. 1. The primers are located by the operator on the dial A, fifty or more at a time, and the cases are located on the dial B as illustrated. Both the dials are rotated, dial A being driven at a constant speed, while dial B is indexed by means of a ratchet dial in the usual manner. Over dial A is a spring that constantly vibrates, due to the friction of the dial rotating under it, so that the primers are agitated and gradually pass into a narrow channel, one at a . time. As the case reaches the priming point, a primer is carried out by a finger, held in line with the pocket in the case, and then a punch operated from a camshaft forces the primer into the pocket. At the same time that the primer is being inserted, the cam on shaft C operates plunger D, which through a fulcrumed lever actuates a padded punch E that holds the cases down while the primer is being inserted. After the primer is inserted, a finger knocks the case from the dial down the chute F. From this chute, it is deposited on a dial G which is also of the ratchet type.

The dial G is used as a medium for holding the cases while the mouth and primer end is lacquered or water-

Fig. 1. Priming and Waterproofing 0.30-caliber Cartridge Cases

proofed. The lacquer is held in tanks H and I, one tank supplying lacquer to a swab that enters into the mouth of the case, while the other supplies lacquer to a swab that lacquers the primer and head of the case. These swabs, of course, are operated in opposite directions, one going into the mouth of the case and the other coming up against the head. Then as the dial indexes around, a ribbon J rotated on pulleys passes across the head of the case and wipes the surplus lacquer from the head. The case then drops out of the dial and is deposited in a box under the machine. The lacquer used for waterproofing the mouth and head of the case is composed of shellac cut with alcohol and resin. The cartridge case is waterproofed so that it cannot become non-explosive if it should be dropped into water. The shellac that is applied is distributed around the rim of the primer and provides a protective coating. The lac-

Fig. 2. Inserting the Powder and Bullets and Crimping—
Loading 0.30-caliber Cartridges

quer deposited in the mouth of the case after the nickel-jacketed bullet has been inserted serves the same purpose.

Loading. — After the primers have been inserted in the cartridge cases, they are ready for loading. This consists in inserting the correct charge of powder and the bullet in the case, and then crimping the bullet into place. The cases are held in the box *A*, Fig. 2, from which they are removed by the operator and placed on a dial of the plain type. This

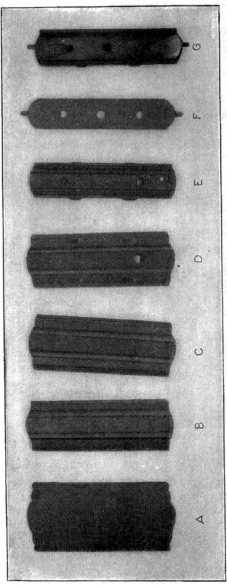

Fig. 3. Evolution of the Cartridge Clip that is used for Holding Five 0.30-caliber Cartridges

dial carries the cases to the ratchet dial B, which, in turn, is indexed and presents the cases mouth up to the various loading containers and punches. The first action on the case is to set it down properly. This is accomplished by a spring punch. Then the proper powder charge is inserted. The powder is held in the funnel-shaped container C and is removed from it by a slide operated by a crank motion. This slide comprises a small container that carries the correct charge of powder and deposits it in the case in which it is packed by means of a punch. Then, as the case indexes around to the

next position, a second charge of powder is inserted. At
the time that the operator is placing cases on the plain dial,
he is also placing nickel-jacketed bullets in the dial D with
the points up. This dial is also of the ratchet type and,
after the powder has been put into the case, it is transferred from dial B to a position under dial D. Then a
punch operated from a camshaft under the machine comes
down on top of the bullet and inserts it in the case. As
the case indexes to the next position, a crimping device
turns in the top edge, holding the bullet in position. It is
then ejected from the machine loaded.

Fig. 4. The Cartridge Clip Machine—A Special Machine that
completes the Body of the Clip in Six Distinct Operations

Making Cartridge Clips. — The device for quickly inserting cartridges in the magazine of a 0.30-caliber rifle
consists of a clip * which holds five cartridges sufficiently
tight to prevent them from falling out. As soon as the clip
is placed over the breech and the top cartridge pressed, they
are ejected and pass into the magazine. The clips are
thrown away when empty, so that they must be made very
cheaply. The main body of the clip is made from a sheet
of brass stock about 0.021 inch thick by 2 7/16 inches
wide. The sequence of operations necessary to complete

* The term "clip" has been used here, as that is the name used by the U. S.
Army; this part, however, should be properly called a "charger." See page 8.

this clip is illustrated from *A* to *G* in Fig. 3, and the machine for making the body of the clip is shown in Figs. 4 and 5.

Referring to Fig. 4, which shows a front view of the press, the strip stock is held on a roll located at the right-hand end of the machine. The stock is fed into the machine by the ordinary type of feeding rolls, and the first operation is to cut out a blank as illustrated at *A* in Fig. 3. This is accomplished by the punch and die *B*, see Fig. 5. The blank is then ejected from the die and carried on to the next punch and die *D* and *E* by means of a transfer slide *C* similar to

Fig. 5.   A Close View of the Dies and Punches used in the
Cartridge Clip Machine shown in Fig. 4

that employed in a multiple plunger press, that receives its motion from a crank mechanism at the left-hand end of the machine. The next operation, performed by the punch *D* and die *E*, is to form two ribs in the center of the blank, and turn up the two edges as shown at *B* in Fig. 3. The formed blank is then ejected and the transfer arrangement carries it on to the next operation, where punch and die *F* and *G* crimp the outer edges into the shape shown at *C*, Fig. 3. The edges of the blank are flattened down and at the same time turned up a distance about 3/64 inch above the top surface of the blank. The blank is then ejected from the die and is carried forward over die *H*. As punch *I* descends it

forces the blank out of the transfer fingers and into the die
H. This operation forms four projections which are shown
at *D* in Fig. 3 that act as retainers for the spring to be inserted later. The next punch and die *J* and *K* draw up the

Fig. 6. Machine used for Making the Springs used in
Cartridge Clips

sides of the clip into a box shape as at *E*, Fig. 3. Then, as
the blank is passed on to the last punch *L* and its die,
it is slightly curved and is ejected from the machine by a
crank mechanism *N*, Fig. 4, which actuates the last die.
This sequence of operations is carried on entirely automati-

cally, and, in fact, the machine will run without any attention whatever until the roll has been exhausted. The operator then starts up the machine and the sequence of operations continues.

Making the Spring for the Cartridge Clip. — In order to hold the cartridge in place in the clip, a curved flat spring F, Fig. 3, is used. This spring, which is made from a sheet of half hard brass stock 0.510 inch wide by 0.010 inch thick, is blanked out and bent to shape in the press shown in Fig. 6. The stock is held on a roll A shown to the left of the machine, and is drawn in by a pair of ordinary feed-

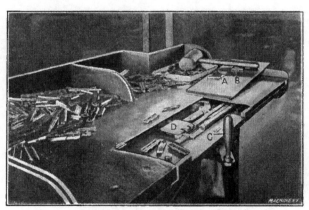

Fig. 7.   Assembling the Spring in the Cartridge Clip

ing rolls B operated by a ratchet mechanism receiving power from the crankshaft of the machine. The first operation is to cut off a strip to the required length, form the ends, and pierce the three holes. This blank, by means of a carrier, is then transferred to the punch and die C. Here the blank is bent up into a curved shape and the spring prongs at each end are formed, after which it is ejected. These prongs are used in assembling the clips for holding the spring in place; they catch on a projection formed in the base of the clip. This machine is also entirely automatic in its operation, and, when once started, it will run until the roll of stock has been exhausted.

Assembling the Spring in the Cartridge Clip. — The assembling of the springs in the cartridge clip is accomplished in the small bench machine shown in Fig. 7. The operator places the clip in a nest, then inserts the spring in a slide which carries it forward and inserts it in the clip. The spring is held on this slide and is pushed into the clip automatically by the prongs which fit into the raised catches in the clip. The clip is carried forward into the assembling position by another slide, which works beside the spring-inserting plunger, and operates a carrier D. In order to show the working mechanism of this machine, the top lid

Fig. 8. Inspecting and Weighing 0.30-caliber Cartridges

or plate that covers the mechanism has been removed, and is shown to the right of the illustration. The clip is inserted through the hole A, and the spring in the hole B. When the operator pulls the handle C, the slide advances carrying the spring and assembles it in the clip. After assembling, the clip is ejected from the machine and drops into a box placed beneath it. The assembled clip appears at G in Fig. 3.

Gaging, Weighing, and Inspecting Loaded Cartridges. — Before locating the cartridges in the clip, they are inspected, gaged, and weighed. These three operations are all accomplished in one machine, which is shown in Fig. 8.

The dies held in the dial *A* in which the cartridges are placed by the operator act as a gage for the body of the cartridge; that is, the contour of the holes in these dies is similar to the chamber of the rifle. As the dial passes around, the cartridges are carried beneath an electrically operated plunger which inspects them to see that each one has a primer in it. Should a cartridge be encountered that has no primer, this punch drops down into the pocket and breaks the electric circuit, which causes a bell to ring, thus notifying the operator that a cartridge with no primer has passed. When the primer is located upside down, the same

Fig. 9.   Cartridge Clipping Machine—Assembling the Cartridges in the Clip

action takes place. The cartridge is weighed in a unique and interesting manner. As the dial *A* passes around, the cartridge is lifted up and caught by the ejector *B*. This transfers the cartridge to the scoops *C* which are carried on the weighing dial *D*. The weighing is done by balances *E* in which the cartridges are deposited by the scoops *C*. The bullet comes up to a stop in these balances and a wire hook attached to the balance catches on the wire *F* when the cartridge carries the correct charge of powder and dumps the cartridge into the box *G*. When the charge of powder in the cartridge is light, the hook on the weighing balance *E* rides up over wire *F*, but as the dial passes around still further the hook catches on a wire located higher than wire *F*, and dumps the cartridge into the light charge box. It is therefore evident that cartridges which are light in weight pass the first box, but cannot go completely around, as they are ejected by the second wire, thus making certain of dumping the weighing balance and throwing the cartridge out. This mechanism successfully eliminates all light charge cartridges, and keeps them uniform in shooting quality.

Inserting Cartridges in the Clips. — The machine used for inserting the cartridges in the clips is shown in Figs.

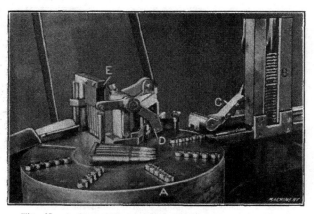

Fig. 10. A Closer View of the Machine shown in Fig. 9,
Illustrating the Operating Mechanism

9 and 10. A dial *A*, which accommodates five cartridges in a row, carries the cartridges around to where the clip is inserted over them. This dial is rotated by means of a crank motion, pawl and ratchet in the ordinary manner, the ratchet dial being located beneath the dial carrying the cartridges. The assembled clips are placed in the proper position in the magazine *B* by an operator; two operators are engaged in keeping the cartridge dial full. The clip is carried out from the bottom of the magazine by means of a carrier operated by an eccentric shaft, and is located over the five cartridges in the dial *A*. The "latch" *C* which is

shown thrown back in the illustration runs under a roller held in bellcrank *D* and seats the clip properly on the heads of the cartridges.   The dial then indexes to the next position, where a bending tool comes into action and bends down the prong projections on the ends of the spring in the clip, thus preventing the cartridges from dropping out. As the dial then indexes around to the next position, the cartridges that have been inserted in the clips are picked up out of the dial by means of a swinging transferring arm that drops them in a box; 65,000 of these cartridges are inserted in the clips per day of eight hours.   Fig. 10 shows a closer view of the machine illustrated in Fig. 9, and gives a better idea of its construction and operation.   Here it will be seen that the ejecting or work-removing fixture is composed of two pieces of sheet steel, spring tempered, which grip the clip by the lower surface.

CHAPTER IX

MANUFACTURE OF SHOT SHELLS

THE ordinary shot shell consists simply of a sheet of paper formed into a tube of the required length, capped with a brass head holding a detonating agent, and then filled with powder and shot separated by wads. The processes in the manufacture of shot shells used by the Dominion Cartridge Co. are outlined in this chapter. The shot shell to be described is shown in its initial stage in

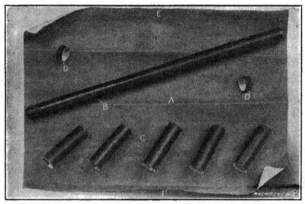

Fig. 1. Initial Stages through which a Shot Shell passes

Fig. 1, where A is the paper sheet 0.0075 inch thick by 14 5/16 by 10¾ inches; B is the sheet rolled into a tube; C is the tube cut into shell lengths; and D the trimmings. The sheet A is made thinner at the edges E, being 0.005 inch thick, so that, when rolled into a tube, a smooth joint will be formed. The component parts of the shot shell dissected and assembled are shown in Fig. 2, where F shows the interior arrangement of the loaded cartridge.

Rolling and Drying the Tubes. — The first operation in the manufacture of a shot shell is the forming of the sheet of paper into a tube, as shown at A and B in Fig. 1. This

rolling operation is carried out in the machine shown in
Fig. 3. A clear idea of the mechanism of this machine.
cannot be obtained from the illustration as it is of a rather
complicated construction, but the principle on which it
works is as follows: The paper sheets, which are in pack
form about 12 inches deep, are placed in a rack held on a
carriage located at the rear end of the machine. The carriage on which the packs of paper are located is raised by
a screw, which, in turn, is operated by a feeding mechanism.
This feeding mechanism is so set that when a number of
sheets have been withdrawn, a finger held on a horizontal
shaft connected with the raising mechanism is tripped, and
the raising mechanism operates until the carrier is raised

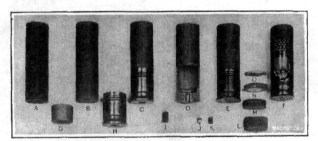

Fig. 2. The Component Parts of a Shot Shell before and
after Assembling

to a sufficient height to continue the feeding operation. The
sheets are withdrawn by two small rollers which travel over
the pack, pick one sheet off at a time and place it on two
rubber belts rotating on four pulleys. These belts carry the
sheet from the pack to the table A located in front of the
pasting and carrying rolls.

In withdrawing the sheets from the pack, they are separated from each other by a gaging mechanism, which allows
only one sheet to pass through at a time. Supposing, however, that this gaging mechanism is not set correctly and
that more than one sheet passes through; then, when the
sheets are carried to the table A, they will not pass under
the arm B, as this arm, if raised to a height greater than
the thickness of one sheet, will stop the feeding mechanism.

The sheet is fed to a stop C fastened to the table, and when located in this position is carried from the table to the carrying roll beneath the pasting roll D, by two rubber belts which also rotate on pulleys, and are located at right angles to the belts that carry the sheet from the pack to the table. The pasting roll is corrugated, so that the paste will run freely along it and be distributed evenly over the sheet. The paste is pumped from a tank E through a pipe F to the pasting rolls, a pump being supplied for this purpose. The pasting and carrying rolls draw in the sheet, and, at the same time, spread the paste over it. As the sheet passes through, it travels on a rack on its way to the rolling arbors.

The method of rolling the tube can be seen by referring to the diagrammatic view, Fig. 4. Here A is the pasting roller and B the carrying roll. The roll B is cut away so that the circumferential distance from C to D equals the width of the sheet.

Fig. 3. Automatic Tube Rolling Machine

This provision is made so that another sheet cannot be drawn in until the one on the arbor is rolled and the arbor indexed. The carrying roll and rolling arbor mechanism are so timed that the rolling arbor makes a sufficient number of turns to complete the tube and index before another sheet is drawn in. The sheet is drawn in over the grate E, which consists of brass strips about 1/16 inch thick and $1\frac{3}{4}$ inch wide, located beneath the rolling arbor when in position for rolling the tube. These rolling arbors *F* and *G* govern the inside diameter of the tube. When the paper is located on the grate, it is driven forward by the carrying roll *B* and comes in contact with the rolling arbor before it leaves the rolls. As it passes the vertical center-line of the arbor *F*, the flappers *H*, which are rubber strips fastened in a roller *I* and rotating in the direction of the arrow, flap the paper up against the rolling arbor. The rolling arbor, at the same time, is driven in the direction indicated, and carries the paper around, thus forming it

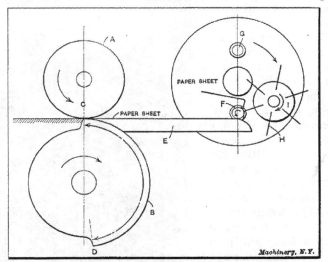

Fig. 4. Diagram showing how the Tube is Rolled

into a tube. As soon as the tube is completed, the indexing mechanism is operated, the arbor *F* being carried to the top and the arbor *G* brought down in position for rolling.

The tube which has been rolled is now located on the arbor and has to be removed. This is accomplished by a bushing located on the rolling arbor back of the tube, and connected to a rack by a short arm. This rack is driven by a spur gear, which, when operated, moves it along, thus carrying the bushing and stripping the tube from the arbor. While one tube is being stripped from the rolling arbor, another tube is being rolled, so that there is a continuous rolling operation. The indexing mechanism is operated by a cam *G*, Fig. 3, which through a lever *H* having teeth cut in its upper end, meshing with a gear (at the rear of the ratchet wheel *L*) on the arbor *I*, rotates the disk *J* carrying the two rolling arbors. The rolling arbors are located in the correct position by means of the pawl *K* and ratchet wheel *L*. The method of driving the rolling arbor when in position for rolling is more clearly shown in the automatic tube cutting machine, Fig. 5.

The tube as it comes from the rolling machine is wet and in a plastic condition, due to the moisture of the paste, thus making it necessary to dry the tube. This is accomplished by putting the paper tubes in a wire basket, and placing the basket in a cupboard heated by steam. This dries the paste very quickly and produces a shell which is extremely stiff, considering the material from which it is made.

Fig. 5. Automatic Tube Cutting Machine

Sizing and Waterproofing. — When the tubes have dried sufficiently, they are ready for sizing. This operation is accomplished by passing the tube through a die which is rotated at a high rate of speed. Lard oil is furnished to the die by a pipe, so that the tube is well lubricated and stripping is avoided. The use of lard oil also extends the life of the die. As one tube is being sized, that is, being forced through the die on the sizing arbor, another tube is being located on the empty sizing arbor. The sizing machine, shown in Fig. 6, is automatic in its operation; all that is necessary in the way of operating is to place the tubes in the hopper *A* parallel with the spindle of the machine. From this hopper they pass down an inclined slide into a pocket. The tube is held in this pocket as it comes down the slide, and is stopped at the rear end, so that it cannot be forced out of the pocket when the arbor is being inserted in it. The tube is taken from the pocket by the empty sizing arbor at the same time that the arbor holding

Fig. 6. Automatic Machine for Sizing the Tubes to the Correct Diameter

a tube is being forced through the die, which is held in the head *H*. In the illustration, the slide *B* carrying the sizing arbors *C* and *D* has retreated from the die, and is in position for the arbors to be indexed.

The tubes, before sizing, are slightly larger than the finished size. The die reduces them to a diameter slightly smaller than the required size, but, after they have been forced through the die, they expand somewhat, so that they can be stripped from the arbor. This machine is provided with a hollow spindle, so that the tubes pass through it and drop into a box. The arbor indexing mechanism is operated through a lever *E* actuated by a cam *F* held on the rear driving shaft *G*. The carriage carrying the arbors *C* and *D* is advanced by a large bevel gear *L* meshing with a rack attached to the arbor slide. The bevel gear *L* is driven by the main driving shaft *G* through bevel gears *I* and *J*.

After the shells are brought to the correct size they are waterproofed, so that they will resist moisture, which is necessary to preserve the powder; as those familiar with cartridges know, powder will not explode when in a wet condition. The water-

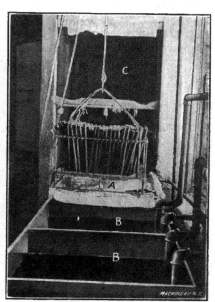

proofing operation consists in putting the tubes into a wire basket *A*, shown in Fig. 7, and then immersing this basket with the tubes in it in one of the tanks *B*. These tanks contain paraffine wax, which is heated and kept in a liquid state by steam pipes passing into the tanks. The tubes are kept in this water-proofing solution until coated thoroughly. The basket is then removed from the tanks and put into the cup-board shown at *C* at the rear of the tanks,

Fig. 7.   Waterproofing Tanks and Drying Cupboards

in which steam pipes are located for drying the tubes.

Cutting to Length and Inserting the Base Wad. — The next operation to be performed on the tubes, after sizing and drying, is to cut them to the length required for the shell. This is done in the automatic tube cutter shown in Fig. 5. The method of putting the tubes on the arbors in this machine is similar to that described in regard to the sizing machine shown in Fig. 6. The tubes are put into the hopper A, from which they come down a slide into a pocket, where they are located in position for the arbor to enter them. The ejecting mechanism used on this machine is the same as that used in the tube rolling machine shown in Fig. 3. Here the ejecting mechanism is more clearly shown. The gear B, driven from the main driving shaft through bevel gears, meshes with a rack C which slides on an arbor D. The ejecting finger E also slides on the arbor D and is connected to the rack. The lower extremity of this finger is bent, and fits in grooves formed on the ejecting bushings F and G, when rolling arbors H and I are in position for stripping the cut tube. The method of driving the cutting

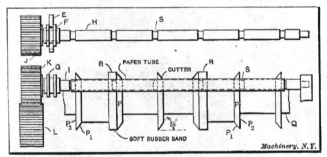

Fig. 8. Diagram showing how the Paper Tube is cut to the Desired Lengths

arbors on which the tube is held is also similar to the mechanism used for rotating the rolling arbors in the tube rolling machine shown in Fig. 3. The arbors H and I have spur gears J and K fastened to them, which, when the arbor is in position for cutting the tube, mesh with gear L. This gear, in turn, is driven by the main driving shaft through spur gears which give it the correct speed.

The method of holding the paper tube on the arbor while cutting is shown in Fig. 8. When the arbor is rotated to the cutting position, the shaft Q carrying the tube cutters P is gradually advanced toward the tube by means of a cam. As the shaft Q advances, the soft-rubber bands R come in contact with the tube, acting as a driver, and thus prevent-

ing the tube from rotating on the arbor or moving longi-
tudinally along it. The shaft Q rotates at the same speed
as the cutting arbors H and I—when in the cutting position.
The cutting arbors H and I are provided with grooves S,
so that the cutters P will not come in contact with the arbor
when they pass through the walls of the tube. The cutters
P are plain disks made from tool steel, hardened and ground
on the faces P_1. It will be noticed from a study of the illustration that the cutters are so located that the faces P_1 are
in line with the edge of the grooves S in the arbor, but are
not facing the shoulder. The reason for this is that as it is

Fig. 9. Machine for Winding and Inserting the Base
Wad in the Shell

necessary to bevel one end of the shell, to facilitate the putting on of the brass cap, the beveled edge P_1 on the cutter
presses the tube against the arbor and acts as a beveling
tool.

Referring again to Fig. 5, the ejecting and feeding mechanism is operated by the face-cam M through levers N and
O. While a tube is being cut, another tube is being placed
on the empty arbor. This machine is entirely automatic in
its operation and cuts the long tube into five different
pieces, and also trims the ends, leaving the scrap shown at
D in Fig. 1.

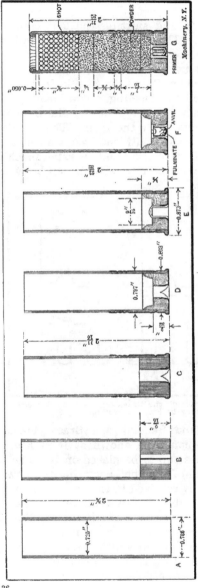

Fig. 10. Transition from the Cut Tube to the Finished Cartridge

Now that the tubes are cut into shells of the desired length, they are ready for the insertion of the base wad shown at G, in Fig. 2, which forms the head of the shell. This operation is accomplished in the hand-operated machine shown in Fig. 9. The paper for the heads is cut into strips of the correct width in another machine. These strips, which are in the form of a roll, are put in the circular box A which has a lid on top of it. The strip is brought out of the slot B in the box, and inserted in a slot in an arbor enclosed in the outer chuck C. This arbor is slightly smaller than the diameter of the battery pocket. When the strip is inserted in the arbor, the machine is started

126

by operating the lever *D*, and shifting the belt from the
loose to the tight pulley. The strip is wound in the bushing *C*, which is of the correct size, and when this bushing
is filled the strip is cut by the sharp edge *E*. The shell
F is now placed in the holder *G* and the handle *H* operated,
carrying the shell toward the bushing. The bushing *C* is
forced back by the carrier *G*, thus exposing the base wad
and allowing it to be inserted into the shell. The handle *H*
is again operated and the shell with the base wad inserted
in it is withdrawn. While this is a hand-operated machine,
it is marvelous how quickly these base wads can be wound
and inserted in the shell.

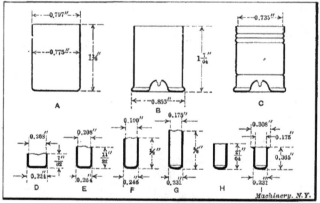

Fig. 11. Stages through which the Brass Cap and the Battery
Pocket pass before Assembling

Expanding, Piercing, and Beading the Brass Heads. —
The shell is now in the condition shown at *B* in Fig. 10,
and is ready for the brass cap to be placed on it. The cap
which covers that portion of the shell enclosing the powder
is made from brass, and is received in the form of a cup as
shown at *A*, Fig. 11. No drawing or trimming operations
are performed on these caps previous to the expanding operation, as they have already been drawn and trimmed to
the exact length. The machine in which these caps are expanded is shown in Fig. 12. The operator places them
mouth up in bushings *A*, twenty-four of which are held in

the ratchet dial B. As the caps pass successively under the expanding punch C, they are expanded. This punch consists of an outer shell which is split and held on a tapered punch by two coil springs wound around it. The expanding punch shown at C is what is called the "starting" punch. Another punch on the same principle, but not shown, is held in the ram of the press, which finishes the expanding of the head. The expanding operation is accomplished by the outer split sleeve sliding up on the tapered p u n c h, which increases its diameter, so that t h e lower end, which is of the desired shape, is enlarged a n d forces out the sides of the head. The bottom of the bushings A are enlarged in diameter to allow the cap to be expanded.

The cap now passes to the punch D which enters it, flattens the h e a d, performs the stamp i ng operation, and pierces the hole, forming a seat for the battery pocket. The hole in this head is not p i e r c e d but is

Fig. 12. Expanding Machine for Partially Forming the Head on the Brass Cap

forced up, no die being used for cutting the material away. The shape of the cap after this operation is shown at B in Fig. 11. The stamping is done by a die having raised letters, so that the impressions are sunk into the head of the shell. This stamp is made with a hole in it, in which the piercing pin is located. As the cap passes around to the punch E, it is flattened, and on further movement, the knockout pin F forces it out of the bushings into a chute, from which it drops into a box. After expanding the heads, the caps are taken to the beading machine shown in Fig. 13, which is of the semi-automatic feed type. The operator places the caps in the slide A with the mouth facing the head of the machine. As they come down this slide, they are located in the pocket, from which they are carried onto a rotating arbor B by a half-bushing held on the shaft C. The beading is accomplished by a circular tool D which is not rotated, but is held stationary to the carrier E. This carrier advances when the cap has been located on the arbor, and the tool D forms the beading. The condition of

Fig. 13. Machine for Beading the Brass Caps

the cap after the beading operation is shown at H in Fig. 2 and C in Fig. 11.

Assembling the Brass Caps and the Paper Shells. — The brass caps and the paper shells are now transferred to the assembling machine shown in Fig. 14, where the cap is placed on the paper shell. Here the paper shells are placed on pins A, sixteen of which are held in the ratchet dial B. Dies C, which are a good fit for the shell, are also held in this dial. The brass caps are placed mouth up on a friction dial located at the rear of the machine by another operator. This machine requires two operators to look after it, although one operator can feed the caps to two or three ma-

chines. As the paper shells are placed on the pins *A* in
the dial *B*, they pass successively under the punches *D*, *E*, *F*,
and *I*. The spring punch *D* seats them partially on the
pins, while the punch *E* seats them correctly. As the shell
comes beneath the punch *F*, the cap is placed on it. The cap
is removed from the dial at the rear, by means of fingers
held on the plate or carrier *G*. This carrier is held on the
shaft *H* which is rotated by a friction gear, driven by a

rack held to the ram
of the press. The plate
G swings through an
angle of about 180 de-
grees, and lifts the
cap from the dial by
means of the fingers,
transferring it to a
position over the shell,
where it is seated by
means of the punch *F*.
The punch *I* seats the
cap firmly on the shell,
and, on the continued
movement of the dial,
the shell is transfer-
red by the pick-up *J*
from the pins to the
chute *K* through the
hollow spindle *L*. The
pick-up mechanism is
held rigidly to the ram
of the press and trav-

Fig. 14. Machine for Assembling the Brass
Cap and Paper Shell

els up and down with it. The condition of the shell after
this operation is shown at *C* in Figs. 2 and 10.

Heading. — In the machine shown in Fig. 14, the brass
cap and shell are only assembled but are not flattened;
neither is the base wad formed to the correct shape. These
operations are carried out in the heading machine shown
in Fig. 15. Here the shells are placed on pins, fourteen of

which are held in the ratchet dial *A*. The dies *B* are also held in this dial, for forming the brass caps. As the dial is rotated, the shells pass successively under the punches *C* and *D*. The punch *C* seats the shell down on the pin, and the punch *D* forces it into the die *B*, flattening the head and forming the base wad. The punches which are held in the dial *A* are made to form the base wad into the shape shown at *D* in Fig. 10. This shape is given to the wad when in a dry condition. The shells are removed by the pick-up *E* which passes over the head of the shell, after it has been lifted from the die by the wad-forming pin. The knockout motion is actuated by the rod *F* held to the ram of the press. This rod *F* operates a mechanism which forces the wad-forming pin up, thus raising the shell out of the dial to a sufficient height for the pick-up to grip it. This pick-up is operated in a manner similar to that use on the verifying machine, Fig. 10,

Fig. 15. Heading Machine in which the Base Wad and Brass Cap are formed to the Correct Shape

Chapter IV, for center-fire cartridge making.

Drawing, Trimming, and Flanging the Battery Pocket. — The battery pocket which is used in the No. 12 gage Imperial shot shell is received in the form of a cup as shown at *D* in Fig. 11. This cup passes through three drawing operations as shown at *E*, *F*, and *G*, which increases its length to about ⅝ inch, and reduces its diameter from 0.324 to 0.231 inch. A drawing press similar to that shown in Fig.

4, Chapter III, is used for the drawing operation, and this cup also passes through the annealing, washing, and drying operations. When drawn to the required length, as at *G*, Fig. 11, the shell is taken to the trimming machine shown in Fig. 7, Chapter III, where it is trimmed to the desired length as shown at *H*.

The next operation is to flange the mouth, as shown at *I*, so that the pocket will have a head which will seat properly in the brass cap of the shell. The flanging operation is performed in an automatic machine which is shown in Fig. 16. This machine is equipped with an automatic feeding device which is on the same principle as that used on the trimming machine shown in Fig. 7, Chapter III. The battery pockets pass from the hopper *A* through a tube located at the rear of the machine to the carriage *B*. This carriage is actuated by levers and cams from the main crankshaft, and carries the pocket from the tube to the flanging dies where the flanging is accomplished by the punch *C*, held in the ram of the machine. An ejecting mechanism operated by the lever *D* ejects the pocket out of the die when it has been flanged, and deposits it in a box.

Fig. 16. Automatic Machine for Flanging the Battery Pocket

Inserting the Battery Pocket in the Brass Cap. — When the battery pocket is flanged, the next operation is to insert it in the brass cap. This is accomplished in the assembling machine shown in Fig. 17. The shells are placed on the pins A, held in the ratchet dial B, and pass successively under the punches C, D, and E. The punch C forces the shells down on the pins, while the punch D seats them down correctly. The punch E sizes the hole in the head of the shell, thus making it ready for the insertion of the battery pocket. The battery pockets are fed automatically to this machine by a hopper-feeding device which is attached to the rear of the ma-

chine. The pockets pass from the hopper to a slide where they drop out. A groove is cut in this slide which is equal to the outside diameter of the pocket, thus making it impossible for the battery pockets to be located in this slide except with the flange up. As the battery pockets are located in the slide, they pass down to the feeding device, which consists mainly of two fingers. One finger carries the pocket from the slide to the second finger which carries it out

Fig. 17. Machine for Inserting the Battery Pocket

and holds it central with the hole in the head of the shell. These fingers are operated by a lever, which is connected eccentrically to the driving shaft. As the finger carries the pocket out and holds it central over the hole in the shell, the ram descends, and a punch held in the bolster K forces the pocket out of the finger into the shell. The pick-up L removes the shells from the pins, and transfers them up through the hollow spindle M, depositing them in the chute

N, from which they pass into a box.   The condition of the
shell after this operation is shown at *E,* Fig. 10.

Automatic Feed Mechanism for Small Brass Cups. —
Another interesting feeding device used in connection with
the hopper on a machine of the type shown in Fig. 17 is
shown in Fig. 18.   This arrangement was designed with the
object of preventing broken tools and of rejecting all cups
that had not previously been trimmed to length.   The cups

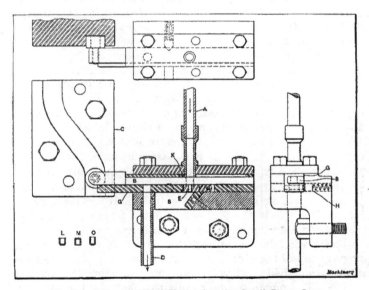

Fig. 18.   Automatic Feed Mechanism for Small Brass Cups

feed down through the supply pipe *A* from the usual feed
hopper, in the position shown at *L.*   Slide *B* is reciprocated
by the plate cam *C* carried on the cross-head.   The cup is
picked up by the cavity in slide *B* and is carried over and
dropped into the supply pipe *D,* from which it passes to the
feed dial.   It sometimes happens that the cup will find its
way into the feed-pipe *A* in an inverted position, as indicated at *M.*   When it enters slide *B* in this position it falls
into the cavity around pin *E* and slide *G* is carried over

along with slide *B*. The function of pin *E* is not primarily to catch the cups that fall with the mouth downward, but to hold up the cups that fall with the head downward, so that the bottom of the cups is level with the bottom of the slide *B*, under normal conditions. When, however, the cup falls with the mouth downward over pin *E*, which is riveted in hinge *F*, the latter drops down over a hardened block, as slide *G* is carried forward, and the inverted cup drops out through cavity *S*. Slide *B* then returns slide *G* to its former position during the return stroke. Normally, the slide is held in position by plunger *H*. A cup that is too long or that has not been previously trimmed, as at *O*, is caught by the edge *K* of slide *G;* the slide is then carried forward and the cup dropped out through cavity *S*, as before.

Piercing the Battery Pocket, Priming, and Inspecting. — The next operation on the cartridge is the piercing of the battery pocket and the inserting of the primer. This operation is performed in the machine shown in Fig. 19. The shells are placed on the pins *A*, sixteen of which are held in the ratchet dial *B*. As the dial passes around, the punch *C* seats the shell down on the pin. This punch also flattens the head slightly and upsets the pocket in the base wad, as shown at *D* in Fig. 2 and at *F* in Fig. 10. The punch *D* pierces the hole in the pocket, while the punch *E* acts as an emergency piercing punch, and also forms the inside of the battery pocket to the correct diameter. A plate *F* is held over the top of the pins to prevent the piercing punches from lifting the shells, when sizing and piercing the battery pocket. The punch *G* again seats the shells on the pins, before the primer is inserted.

The primers are placed in the brass tube *H*, which is held in the bushing *I*. This tube passes up at the rear of the machine, and is encased by a sheet-steel plate, so that, if the primers in the tube should explode, the operators would not be injured. The primers drop through this tube and are removed from it by a finger *J* held to a slide, which carries the primer to a central position over the pocket. This slide is actuated through levers connected to the crankshaft of the press. As the finger carrying the primer advances to a central position over the pocket in the shell, the punch K descends and forces the primer out of the finger into the pocket. The dial is then rotated and the pick-up L removes the shell from the pin and transfers it to the chute M, as previously described. This chute is made from wire netting, and leather strips are placed on the side to prevent the shell from being bruised.

The object of having the wire netting is to allow any

loose fulminate, which may have been removed from the primer, to drop into the box N, which is partially filled with water. If this fulminate were allowed to go into the box in which the shells are placed, an explosion might result, so that it is well to take this precaution. The pierced and primed shell is shown at F, Fig. 10. Before the shells can be transferred to the loading department, it is necessary for them to pass through a rigid inspection. This consists in seeing that all primers are located below the top surface of the head of the shell, and also that the paper case is not dented or marked in any way. Any shells which are found to have these defects are scrapped.

Fig. 19.   Machine for Inserting the Primers

Cutting and Greasing the Felt Wads.— As can be seen at G in Fig. 10, three felt wads are used to separate the shot from the powder. These wads are also shown at L, M, and N in Fig. 2, and are used for two reasons: One is to separate the powder from the shot and to give it the required compression; and the other is to prevent the powder from becoming damp. To perform this latter function, it is necessary to grease the wads, so as to make them waterproof. This is done in the machine shown in Fig. 20, in which the wads are also cut. The felt for these wads comes in strips about 10 inches wide and is fed into the machine by the rollers *B*. Fourteen punches *C* are held to the ram of the press for cut-

ting out the wads, and cutting dies registering with these punches are held in the die-block *D*. As the wads are cut out, they pass through the dies into tubes located in the die-block. These tubes have holes drilled through them, and around their peripheries, through which grease held in a reservoir in the die-block can pass in to the wads. The grease is kept at the proper temperature by means of steam pipes, which pass into the die-block. The greasing tubes

Fig. 20. Machine for Cutting and Greasing the Felt Wads

pass through the bolster and down through the bed of the press, so that the wads, as they are cut out and greased, fall into a box under the machine. The three felt wads shown at *L*, *M*, and *N* in Fig. 2 are all cut out and greased in a similar manner in this machine.

Loading and Testing. — All the component parts of the shot shell, which are illustrated in Fig. 2, have been completed. The operation now to be accomplished is the as-

sembling of these various parts in the paper container or
shell, thus making a completed cartridge.   The various
parts as made are transferred to the shot-shell loading department where the shells are placed in the slide of an
automatic loading machine.   The shells pass from this slide
down a chute from which they are located in holes in a
ratchet dial.   As the shells are located in the dial mouth up,
they first pass under a container in which is held a sufficient charge of powder.

The container which holds the charge of powder has a
slide beneath it which is actuated by an eccentric rod connected to a shaft at the rear of the machine.   The function
of this eccentric rod is to open and close the container, so
that the powder can drop out into the shell.   As the shells
pass around further, the wads are deposited in them.   These
wads are held in four brass tubes which hold the various
wads in the correct order in which they are to be placed in
the shell.   The wads are carried from these tubes out to a
position central with the shell by slides which are actuated
by eccentric rods, as already mentioned.   All the wads are
fed to the shell in this same manner.   When the last of the
felt wads is placed in the shell, the shot is inserted.   This
is held in a container, and is removed from it in the same
manner as the powder.   As the dial passes around still
further, the cardboard wad, on which the number of grains
of powder and the size of the shot are marked, is taken
from the last of the four tubes, and punches are used to
seat the wads in the shell as they are carried out by the
finger and held in a central position.

The last operation on the shell is the turning in of the
top edge, or the "crimping operation" as it is called.   This
crimper consists of a die in which hardened pins are driven.
These hardened pins are so formed that they turn over the
top of the shell.   The crimper is driven by a belt, which
runs over two idler pulleys, and drives a pulley which is
connected to the spindle holding the crimping die.   After
the loading and crimping operations the cartridge is in the
condition shown diagrammatically at G, Fig. 10.

This completes the manufacturing operations, and the cartridges are now ready for the last inspection, which consists in testing them for accuracy. As a shot shell does not contain one bullet, but a collection of small shot, the speed of the shot is not a very essential consideration. However, the charge of shot must be confined to a certain space when fired at a specified distance. The No. 12 gage Imperial shot shell is not allowed to pass the testing inspection unless two-thirds of the charge of shot fills a 30-inch circle, when the cartridge is fired at a distance of 30 yards. As the number of grains of shot in the shell is known, it is an

TABLE I. SIZE AND WEIGHT OF "DROP" AND "CHILLED" SHOT

(Tatham Bros. American Standard)

Size of Shot	Diameter in Inches	Number of "Chilled" Shot to the Ounce	Number of "Drop" Shot to the Ounce	Size of Shot	Diameter in Inches	Number of "Chilled" Shot to the Ounce	Number of "Drop" Shot to the Ounce
Dust	0.040	....	4565	No. 4	0.130	136	132
No. 12	0.050	2385	2326	No. 3	0.140	109	106
No. 11	0.060	1880	1346	No. 2	0.150	88	86
10½ Trap	0.065	1130	1056	No. 1	0.160	73	71
No. 10	0.070	868	848	B	0.170	...	59
9½ Trap	0.075	716	688	Air Rifle	0.175	...	55
No. 9	0.080	585	568	B B	0.180	...	50
8½ Trap	0.085	495	472	B B B	0.190	...	42
No. 8	0.090	409	399	T	0.200	...	36
7½ Trap	0.095	345	338	T T	0.210	...	31
No. 7	0.100	299	291	F	0.220	...	27
No. 6	0.110	223	218	F F	0.230	...	24
No. 5	0.120	172	168	....	......	...	...

easy matter to determine whether these shells are loaded correctly or not. Shot shells are packed by hand in boxes which contain 25 and 50 cartridges, respectively.

Shot Charges for Shot Shells. — Shot shells are loaded with different shot charges depending upon the requirements. For duck or fowl shooting, small-sized or "bird" shot is generally used; for big game shooting, single shot loads are generally employed. Shot used in shot shells may be divided into four distinct types: Drop shot, chilled shot, buck shot, and balls. Drop shot is produced by dropping molten lead from a tower through a sieve into water; chilled shot is produced in the same manner, but, instead of using pure lead, an alloy containing lead and antimony is used. Buck shot, especially the larger size, is cast or swaged; so, also, are lead balls. Table I gives the principal sizes in which drop and chilled shot is made, and Table II, the sizes and weights of buck shot and lead balls.

In loading shot shells with small size shot, the charge is measured and depends more upon weight than upon the number of shot used. This is not the case, however, when buck shot or balls are used, because of the necessity of "chambering" the charge. Chambering of the charge refers to the number of shot pellets in each layer of the charge

TABLE II. SIZES AND WEIGHTS OF LEAD BUCK
SHOT AND BALLS

(Tatham Bros. American Standard)

Size of Shot or Ball	Diameter in Inches	Number to the Pound (Approx.)	Size of Shot or Ball	Diameter in Inches	Number to the Pound (Approx.)
4C	0.240	341	Ball	0.430	55
3C	0.250	299	Ball	0.440	50
2C	0.270	238	Ball	0.480	45
1C	0.300	175	Ball	0.500	40
0	0.320	144	Ball	0.530	32
00	0.340	122	Ball	0.580	24
000	0.360	103	Ball	0.640	18
Ball	0.380	85	Ball	0.650	17
Ball	0.400	70	Ball	0.660	16
Ball	0.420	64	Ball	0.680	14

that fill the bore of the gun at the muzzle without overlapping. The gage of a shotgun is derived from the number of shot required to make a pound; thus, the caliber of a 16-gage gun is equal to the diameter of a single spherical shot weighing 16 to the pound (see Table III); 12-gage, 12 shot to the pound, etc.

Manufacture of Drop and Chilled Shot. — As   mentioned, drop and chilled shot is made by dropping molten lead from a tower. This process is used in making shot up to the TT size, and occasionally up to the FF size, which, as shown in Table I, is 0.230 inch in diameter. Difficulty is generally experienced, however, in dropping shot much larger than 3/16 inch in diameter, and shot and balls above this size are either cast or swaged. The method generally employed in dropping shot may be briefly described as follows: The building used for this purpose is known as a "shot tower," and contains a circular sieve frame provided with a pan having round holes of a diameter corresponding to the size of the shot to be dropped. Located near the pan at the top of the tower is the melting pot and furnace. The molten lead is ladled from the pot to the surface of the pan, which is covered with lead dross, so that the molten lead percolates through slowly. Another method consists in siphoning the molten lead onto the sieve. The molten lead, in dropping through the air, is formed into small glo-

TABLE III. BUCK-SHOT CHARGES FOR 20, 16, 12, AND 10 GAGE SHOT SHELLS

(Tatham Bros. American Standard)

Gage of Shot Shell	Size of Shot	Pellets per Layer	Pellets per Load.	Gage of Shot Shell	Size of Shot	Pellets per Layer	Pellets per Load
20	1 C	2	6	12	1 C	4	12
20	2 C	3	12	12	2 C	5	15
20	3 C	4	16	12	3 C	5	20
16	00	2	6	12	4 C	7	27
16	1 C	3	9	10	000	3	9
16	2 C	4	12	10	00	3	9
16	3 C	4	12	10	0	4	12
16	4 C	5	15	10	1 C	4	12
12	000	2	6	10	2 C	5	20
12	00	3	9	10	3 C	7	27
12	0	3	9	10	4 C	7	27

bules and cools sufficiently before striking the water in the well to prevent it from flattening. The distance from the surface of the screen to the surface of the water is from 100 to 200 feet. The lead used must be free from zinc, and must be alloyed with a small percentage of arsenic, otherwise the globular form will not be obtained.

From the well, the shot is conveyed to a steam-jacketed tumbling barrel, where it is dried, and from here it passes to the grading tables. These grading tables are located on an incline, and consist of a series of "steps" with gaps between, each "step" being slightly lower than the preceding one. The spherical shot rolls down the tables and jumps across the gaps. The imperfectly formed shot, "twin" shot, etc., will not roll with sufficient speed, and, consequently, drops through the gaps into scrap boxes; this imperfect shot is again remelted. After grading on the tables, the perfect shot is taken to tumbling barrels where it is mixed with plumbago and tumbled to polish it. From here it is taken to the sizing screens; at this point, the shot is segregated into compartments containing shot of the various sizes, and is then conveyed into the proper bins, from which it is drawn and packed in bags ready for shipment. The production of one shot tower is from fifty to seventy tons of shot in ten hours. All sizes of shot cannot be dropped satisfactorily. The limit is about "TT," as given in Table I. Shot larger than this is cast in molds or swaged and formed similar to bullets. This process is also followed in making lead balls which are sometimes made up to one inch in diameter.

CHAPTER X

MANUFACTURING THE FRENCH MILITARY RIFLE CARTRIDGE

THE processes followed in the manufacture of the French 8-millimeter (0.315 inch) caliber rifle cartridge differ considerably from those employed in the production of other military rifle cartridges. In the first place, it will be noticed upon reference to Fig. 1 that this type of cartridge is so made that it holds a large charge of powder in proportion to its length, and, consequently, is more truly "bottle shaped" than any of the other military cartridges. In fact, with the exception of the German "Mauser" and American "Springfield" cartridges, it contains a larger charge of smokeless powder than any of the others. The bullet also differs from those used by other great powers; instead of being metal-cased, the bullet is made solid from a composition copper rod.

Blanks for the Cartridge Case. — The first operation in the manufacture of the French military rifle cartridge is the production of the blank from which the case is subsequently made. This blank, as shown at *A* in Fig. 2, is 23 millimeters (0.9055 inch) in diameter and is cut out from a sheet that is 150 millimeters (5.9056 inch) wide by 4 millimeters (0.1575 inch) thick. The length of this sheet (which is not in roll form) varies from 1.3 to 2 meters (one meter equals about 3¼ feet). The blanks, which are cut out seven at a time and at the rate of 420 per minute, weigh 13.8 grams each. By referring to *A* in Fig. 2, it will be seen that the blank is slightly curved; this curvature is produced by the punch when the blank is cut out, as shown at *A* in Fig. 3. The composition from which the sheet is made is 67 per cent copper and 33 per cent zinc. The tensile strength is from 30.6 to 34 kilograms per square millimeter. The elongation on a test piece 100 millimeters long is 57 per cent as a minimum. The material must have no flaws, cracks, or spots.

Cupping, Indenting and Redrawing Operations. — Following the cutting out of the blank, as shown at *A* in Fig. 3, the next operation is cupping. This operation is handled in a double-action power press of special design, which is equipped with a magazine feed attachment, but is handled in a manner different from that ordinarily employed in the making of cups for cartridge cases. In this case, the aim is to throw the metal to the head rather than to lengthen the walls. This is accomplished in the manner shown at *B* in Fig. 3. The blank is fed to, and forced into, die *a*, punch *b* thus forming it to the desired shape. Upon the up-stroke of the ram carrying the die, punch *c* ejects the finished cup from the die. The metal is confined between punch *b* and die *a*, and, consequently, is prevented from "flowing up" along the punch, as is usually the case; *B*, Fig. 2, shows the

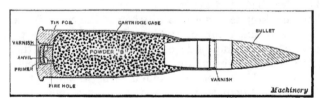

Fig. 1.   Section through French 8-millimeter Military Rifle Cartridge (Scale about Actual Size)

shape and size of the cup after this operation. Here it will be noticed that the largest diameter of the cup is 0.3 millimeter greater than the blank, and that the thickness at the bottom has been increased from 4 to 4.4 millimeters.

Following this operation, the cup is not annealed, but is taken directly to the second cupping operation shown at *C* in Fig. 3. Here the blank, while not confined in the same way as at *B*, is still confined sufficiently to prevent the metal from flowing from the head, and, at the same time, the insert is made in the base of the cup. In this case, the cup is forced up by punch *d* into die *f* and against punch *e*, resulting in the formation of a cup as shown at *C* in Fig. 2. This operation is performed in the same type of press as that used for the first cupping operation.

Without annealing, the cup is now given the first redraw,

as shown at D in Fig. 3. This is accomplished in the same manner as that usually employed for drawing cartridge cases, the cup being forced through the die, as illustrated, in a power press carrying two dies and punches. The cartridge case, still without any annealing operation, is then passed to the second redraw, which is handled in the same manner as that shown at D, Fig. 3, and produces a case of the shape shown at E in Fig. 2.

The case is now annealed. For this operation, the cases are put in revolving drums which are placed in a furnace heated to a temperature of from 500 to 550 degrees C. (about from 930 to 1020 degrees F.). The cases are allowed to remain in the furnace until they attain this temperature —about 40 minutes—when they are taken out and cooled in water. Following the annealing operation, the cases are pickled, washed, and dried.

The next step is to select a number of the cases for analysis to see if the grain of the metal is right. The case is sectioned and a photo-micrograph made. This is then inspected to see if the grain of the metal has the necessary refinement. If the metal is found to be in the right state, the cups are taken to the drawing presses and given the third redraw, producing a case of the shape shown at F in Fig. 2. Following this, the cases are trimmed to the length shown at G. The length of the case varies and is dependent upon the weight. In other words, the length of the case after this operation should be such as to weigh between 11.6 and 11.7 grams. The weight of the finished case is 11.5 grams. The trimming is done in a special horizontal trimming machine provided with a magazine feed. The circular disk cutter used for trimming is revolved in the opposite direction to the work.

Heading, Mouth-annealing and Tapering.— Without any annealing, the cartridge case is now headed, and, owing to the peculiar shape of the head and primer pocket, two heading operations are found necessary. These two heading operations, however, are accomplished at the same time. The machine used is a special power-driven header, equipped with a ratchet dial and carrying two heading punches. It is

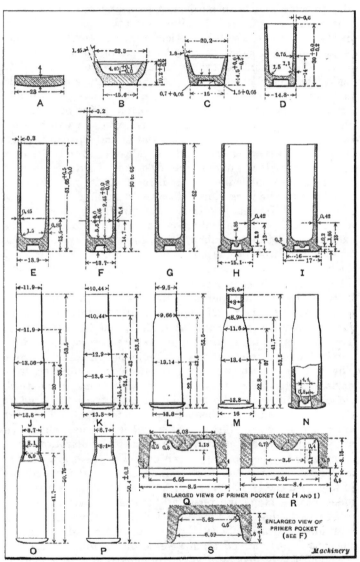

Fig. 2. Sequence of Operations on French Military Rifle Cartridge

also provided with an automatic feed and ejecting device. The result of the first heading operation is shown at *H* in Fig. 2. Here it will be noticed that the anvil projection is started and is made smaller than the finished size shown at *I*. The reason for this is that if it were attempted to make this teat larger in the first operation, a ring would be formed in the second operation, as the excess metal would necessarily be carried down to the bottom of the pocket. The shape of the primer pocket is also changed, as will be seen by referring to the enlarged views at *Q* and *R*, respectively. As the dial carrying the work is indexed around, the case comes successively under the first and then the second punch; the second punch, in addition to flattening the head and forming the pocket, also does the stamping.

Following the heading, the case is partially annealed with a gas torch. The annealing is done on the mouth of the case for a distance of about 20 millimeters down, and the temperature attained is about 600 degrees C. (about 1100 degrees F.), or what is known as a "cherry red" color. After being allowed to cool off in the air, the mouth and body of the case is tapered. Owing to the extreme bottle-neck shape of this cartridge case, four tapering operations are necessary, as illustrated at *J*, *K*, *L*, and *M*, respectively. The first three tapering operations are accomplished in a vertical power press of the multiple punch and die type. The last or fourth tapering operation is performed in a horizontal heading press equipped with a magazine feed. Following tapering, the burr is removed from the head of the cartridge case in a head trimming machine. The next operation is the drilling of the small vent holes, as shown at *N*, which is accomplished in a special horizontal two-spindle high-speed drilling machine. The cases are carried in a turret that automatically indexes around to positions in line with the two drill spindles. The machine is also pro-vided with a magazine feed and automatic ejecting device.

The cartridge case now passes through two sizing and trimming operations. In the first operation, shown at *O*, the length is trimmed from 53.5 to 50.76 millimeters, and the mouth enlarged by a sizing punch from 8 to 8.1 milli-

meters. In the second operation, shown at *P*, the length
is reduced to 50.4 millimeters, and the sizing punch checks
up the hole for size. These two operations are performed
in the same machine, which resembles in design that used
for drilling the vent holes. The machine has two spindles
carrying two sizing punches and trimming tools.

The next operation on the cartridge case consists in annealing the mouth of the case for a distance of from 5 to 6

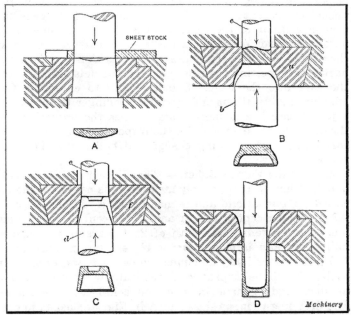

Fig. 3. Diagram illustrating Method of Making Blank and Performing First
and Second Cupping and First Redrawing Operations

millimeters from the mouth end, at a temperature of 600
degrees C. (about 1100 degrees F.). A gas torch is used
for this purpose. This annealing operation softens the case
at the extreme mouth, which, in a subsequent operation, is
curved in to fit the bullet.

Priming and Inspecting. — The case is now washed and
then taken to a priming machine where the primer is in-

serted. This operation is performed in a special machine carrying two dials, one horizontal and the other vertical. The case is held in the vertical slide which is fed by a magazine, and the primers are fed into the horizontal dial in a similar manner. The punch used for seating the primer comes up from the bed of the machine. Upon the completion of the machining operations, the cartridge case passes through several very rigid inspections. One inspection consists in seeing that the primer is placed tightly in the pocket to prevent escape of gas. The other is to gage the correct contour of the cartridge case. This is done in an automatic machine which inspects these cartridge cases at the rate of 30,000 in ten hours. The machine used is of the horizontal type, and a spring-operated punch is used to force the case into the gage. Should it fail to enter the full amount, a slide hits it and operates a plunger to throw the rejected case into a box. This finishes the operations on the cartridge case, except for the inspection operations performed after the loading charge and bullet have been inserted.

Making the French Bullet. — The French military rifle bullet, as has been previously mentioned, is not of the metal-case type, but is solid and is made from a rod of wire containing 90 per cent of copper and 10 per cent of zinc. This must have a tensile strength of 25 kilograms per square millimeter, and an elongation of 42 per cent on a test piece 100 millimeters long, and must show no flaws, cracks, or spots. The bullet is produced practically without any waste of stock and is made in a special cold-heading machine. The first operation, as shown at A in Fig. 4, consists in cutting off a blank of the required length. The length, however, is not an important factor, as the size is governed by the weight, which must be from 13.17 to 13.25 grams. The cutting off is done in a cold header in the manner shown at A in Fig. 5. In the first position, shown at A, copper rod a is fed through die b and is cut off by shearing knife c. After cutting off, the wire is carried over from position A to position B by finger d, and is driven into die e and headed by punch f. Upon the retreat of the punch f, ejector g is

advanced and forces the finished plug out of the die.   The
shape and size of the bullet after this operation is shown at
B in Fig. 4.

Forming and Trimming the Bullet. — The next operation,
which is known as the first compression, is also accomplished

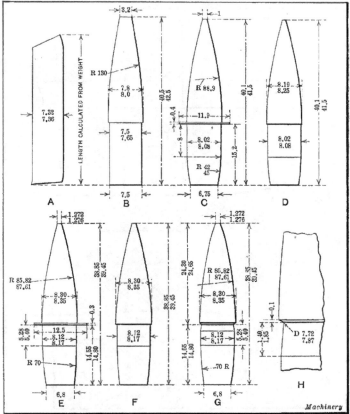

Fig. 4. Sequence of Operations on a French Military Rifle Cartridge Bullet

in a cold-heading machine equipped with a magazine attachment.   The diagram shown at *C* in Fig. 5 illustrates
how this operation is carried out.   The bullet which is fed
down from a slide is inserted in die *h* and then compressed

by means of die *i*. This operation forms a fin, as shown, which is removed in a subsequent operation. Upon the retreat of the slide carrying die *i*, ejector *j* forces the finished blank out of the die. The blank does not stick in die *h*, owing to the shape of the latter, and also to the wedging action

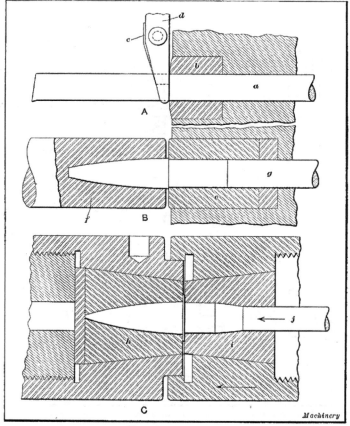

Fig. 5. Diagram Illustrating Method of Cutting Off and Forming the Blank for the Bullet, and Performing the First Forming Operation

produced when the blank is swaged into the die *i*. The friction between the blank and die *i* is greater than that between the blank and die h; consequently, the latter can be withdrawn, leaving the blank still retained in die i.

Following the first compression, the fin that is formed is removed in a horizontal header. This is accomplished by means of an ordinary shearing punch and die, the header being equipped with a magazine feed, and the blank forced through the die. The shape of the bullet after the first compression and shearing operation is shown at C and D in Fig. 4, respectively. The next operation, known as the second compression, is accomplished in a similar manner to that used for the first compression, and the result of this operation is shown at E. The compression of the blank into bullet form in the cold header hardens the metal considerably, so that an annealing operation is necessary. This is performed on the bullet after the second compression, shown at E in Fig. 4. The annealing is done in a furnace in which the bullet is heated to a maximum temperature of 400 degrees C. (about 750 degrees F.). The bullet is now taken to a horizontal header, and the burr formed in heading removed in the manner previously described.

The next machining operation consists in cutting a groove around the periphery of the bullet, which is used for crimping in the mouth of the cartridge case. This groove is cut in a special automatic lathe, where a circular form tool is used for doing the turning. An enlarged view of this groove is shown at H in Fig. 4. The bullets are now examined for outside defects, and ten per cent inspected as to size and weight. The weight of the finished bullet is from 12.65 to 13 grams.

'Loading and Inspecting the Cartridge. — With the exception of the primer, the cartridge case and bullet are now completed and ready for loading. As the manufacture of the primer is covered in a separate chapter, it will not be dealt with here. Upon the completion of the cartridge case and bullet, both members are transferred to the loading department. Here the correct charge of powder in flake form is measured by weight and inserted in the cartridge case. The number of grams of powder used varies, depending on the characteristics of the powder itself; the requirements are that the powder must develop a certain breach pressure and give the bullet a certain muzzle velocity. The breach pressure must be 3000 kilograms maximum, and the muzzle velocity 680 meters per second, minimum. Upon the weighing of the correct charge and the placing of it in the cartridge case, the bullet is inserted, and the mouth of the case crimped to hold the bullet tightly in place.

Following the loading and crimping, the entire cartridge is tested for concentricity in a special machine. The gage used covers both the case and the bullet. Varnish is next deposited in the primer pocket around the primer and at the point where the case is crimped on the bullet; this is done to make the cartridge water-tight. The varnish is tested for moisture by putting the cartridges in water and leaving them there for ten hours. The bullet and primer are then removed and an inspection made to see if the powder is damp. If it is damp, the varnishing has not sufficiently waterproofed the cartridge and the reason for this defect is ascertained before any more cartridges are put through: After the cartridges have finally passed all manufacturing inspections, they are subjected to a firing test. The target used is 2 meters square and is located 200 meters from the muzzle of the rifle. Ten cartridges are selected at random for testing, and after these have been fired at the target, the target is taken down and measured, and no two of the perforations in the target made by the bullet must be more than 500 millimeters apart. If one cartridge out of the ten is outside of this distance, a second test of ten is made. If the cartridges satisfactorily pass this test they are ready for packing.

CHAPTER XI

MANUFACTURE OF PRIMERS

PRIMERS, or percussion caps, as they are sometimes called, are held in the head of the cartridge case and are used to detonate the propelling charge in the case. Primers for use in military cartridges are made in two distinct types, as shown diagramatically in Fig. 1. In one type A, the anvil is formed as an integral part of the primer pocket in the head of the case, whereas, in the other type, shown at B, the anvil is a separate part a and is held in the primer cup proper. In both cases, the detonating charge is pressed into the cup and is then covered with a lacquered sheet of tin foil b. This is used to prevent the detonating charge from falling out of the cup, and also to make the primer waterproof. Primers in which the anvil is a separate part are sometimes made without the tin-foil disk, and lacquer is used instead. The flame produced by the discharge of the primer reaches the propelling charge through small vent holes c. In the diagram shown at A, two vent holes are provided, which are located on opposite sides of the anvil projection. In some cases, the vent hole passes through the center of the anvil projection and the firing pin is located off center. In the type shown at B, only one vent hole, as a rule, is provided, and this is located in the center of the pocket.

An entirely different arrangement of the primer cup and anvil is used in shot shells, which are provided with battery pocket heads, as shown at C. Here the primer cup d is held in a battery pocket e, the latter being held in the head wad f and the thin brass head g. The anvil, in this case, is a flat punching which is arranged parallel with the axis of the pocket and presses against the inner face of the bottom of the pocket e and against the tin-foil disk b.

Blanking and Cupping Primer Cups.— The first step in the manufacture of a primer is the production of the cup.

This is generally cut out from a copper alloy strip known as "gilding metal" and composed of 90 per cent of copper and 10 per cent of zinc. This composition is not universally adopted, and, in some cases, the zinc content is increased with a corresponding decrease in the copper content. The requirements of a primer cup are that it be capable of being detented considerably by the firing pin without being punctured. The firing pin, as a rule, strikes the primer cup with an energy of about 17 inch-pounds. There are two methods of cutting out primer cups. One is to cut out a large number at each stroke of the double-action punch press, the number depending upon the size of the cup and the type of press used. Double-action presses are so equipped that as many as sixteen cups are produced at each

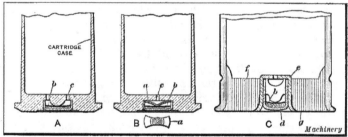

Fig. 1. Diagram Illustrating Different Types of Primers

stroke of the press. Another method consists in cutting out one cup at a time in a much smaller double-action press provided with a roll feed and using strip stock of a width sufficient to make one cup. The production obtained by this method, of course, is much less than that where a number of cups are produced at each stroke of the press, but the small press can be run at a much higher speed than the larger one and also represents a reduced initial cost for installation. The multiple punch type of press is generally operated at from 90 to 100 R. P. M., and the small single punch press, at a speed ranging from 175 to 200 R. P. M.

The diagram shown in Fig. 2 illustrates the method of making primer cups by the last-mentioned process. At *A*, the strip of stock is placed over the combination blanking

and cupping die *a*, and the blanking punch *b* is shown resting on it.   At *B*, the blanking punch has come into operation, has cut out a blank of the required size, and is acting
as a blank-holder.   At *C*, the cupping punch *c* has come into
action on the blank, and is starting to force it through the
die *a*.   At *D*, the cup is completed.

The process just described is that followed in the manufacture of primers used in the ordinary rifle cartridge. The
making of long primers as used in high-grade paper shot
shells requires the additional operations of redrawing and
trimming.   These operations are performed in drawing
presses of the type illustrated in Fig. 4, Chapter III.   The
end trimming is performed in an automatic trimming machine provided with what is called a "fountain feed" type
of magazine.

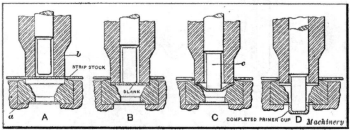

Fig. 2. Sequence of Operations in the Production of a Primer Cup in a
Double-action Punch Press

Blanking and Forming Primer Anvils. — After the primer
cup has been made, as previously described, no other machining operations are performed on it, and, after thorough
cleansing, it is ready for charging.   However, before the
cup can be charged and is ready for use, it is necessary
to make the anvil, when the design of the cartridge case requires this part to be a separate member.   The anvil, which
is made from hard brass, is generally produced in a single-
action press provided with a ratchet roll feed.   Narrow strip
stock is used, which is in the form of a roll.   This stock is
held in a circular box at the front of the machine and drawn
between the tools by means of a roll feed engaging with the
pierced strip.   When a single punch and die is used, the

press is generally operated at from 175 to 200 R. P. M. The diagram, Fig. 3, shows the type of punch and die used for making anvils. The strip stock is pulled over the top of die *a*, as previously described, and punch *b*, in descending, blanks out the anvil *c*, and, at the same time, forms it as shown at *B*. Sometimes, a gang of punches and dies is used; the greatest number recommended is five separate sets of tools. After the anvil has been washed and dried, it is ready for assembling in the primer; this operation, of course, following the charging of the latter.

Charging Primers. — The charging of primers is a delicate and more or less dangerous operation, and requires considerable experience and care to obtain the best results.

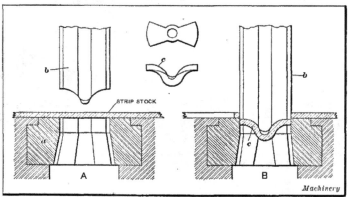

STRIP STOCK

A   B

Machinery

Fig. 3. Method of Blanking and Forming Primer Anvils

There are two principal methods in use. In the first method, the detonating charge is inserted in a dry condition, and, in the second, in a wet condition. When primers are charged by the first-mentioned process, no extended drying is necessary, but, in the second method, the primers must be placed in drying cupboards where they are left for several days. The following description pertains to the charging of primers by the dry method.

The first step is to shake the thoroughly cleansed primer cups into a primer charging plate, shown in Fig. 4. This plate may be made of a size and with the number of holes desired, the holes being counterbored to the exact size and depth demanded by the cups. These plates, after being filled with the primer cups, are placed in a primer varnishing machine. This machine is usually of the bench fixture type, and so designed that the punches, which are carried in a separate member, can be successfully dipped into a varnishing trough, and brought in line with the primer cups held in the plate. The first step is to operate this device so that the punches are dipped into the shellac bath. Then the punches are brought over in line with the cups and deposit a small drop of varnish in each cup. The charging plate with its varnished cup is then taken from the varnish-

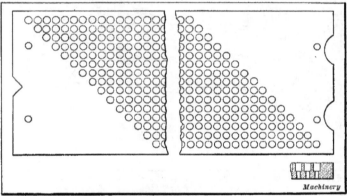

Fig. 4.   Primer Charging Plate designed to hold 600 Primer Cups at One Charging

ing machine and put on a drying table. This table can be heated either by steam or gas and the cups are allowed to remain on it until the varnish becomes adhesive. When the primer charging plate has become sufficiently cool, it is removed to the primer charging table, where the dry fulminate or detonating charge is inserted.

Some idea of the method followed in charging primers may be obtained by referring to Fig. 5. This shows one type of charging table that is used in connection with the dry mixture process. In operation, the charging plate with its varnished cups is put on the sliding shelf A, the exact position being registered by dowel-pins. The shelf is then slid under the charger plate, which consists of two separate bronze plates with the same number of holes as that contained in the primer charging plate, and, of course, located in the same position. The upper one of these two plates is made of an exact thickness to measure the amount of detonating composition that is required for each primer. This upper or top plate is free to slide endwise, by operating a long handle, not shown in the illustration. These two plates must be made with extreme accuracy and so nicely fitted together that no mixture can get in between them.

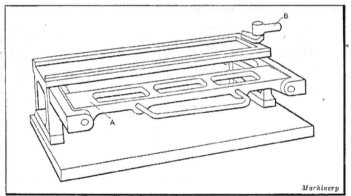

Fig. 5. Charging Table used in Charging Primer Cups by the Dry Mixture Process

The handle B is turned so as to slide the holes in the top plate out of alignment with the holes in the bottom plate and a dry mixture of the detonating composition is poured onto the plate from a soft rubber box. The holes are then filled with the aid of a soft rubber strip or brush attached to a long handle, all of the mixture remaining on the plates being brushed off into the rubber box which is at once removed from the machine to a place of safety. The handle is then turned back, bringing the holes in both plates in line with each other, thus allowing the detonation pellets to drop through the bottom plate into the primer cups in the charging plate beneath it. The slide is then withdrawn and the charging plate taken to the primer press.

Pressing the Detonating Composition.— The press used for compressing the detonating composition in the primer cup is designed somewhat along the principle of a regular punch press, but carries a special table in which the primer charging plate is inserted. This table is so designed that the fixture which carries the charging plate is indexed a distance equal to the space between each consecutive line of holes in the charging plate at every stroke of the press. The press, for instance, that would be used for pressing the powder in primers held in the plate shown in Fig. 4, would carry fifteen punches arranged in a diagonal position, and forty strokes of the press would complete the pressing of the detonating composition in the 600 primers held in this plate. After pressing the last row of primers, the machine automatically stops; the charging plate still holding the charged cups is then returned to the varnishing machine, as previously described, and a final drop of varnish inserted. The cups are then ready to receive the tin-foil disks.

The press used for cutting out and inserting the tin foil in the primer cups is of a similar construction to the press just described, with the exception that the "foiling" press is provided with an automatic roll feed for feeding strips of tin foil over the cutting dies, under which the charging plates are located. The strip of tin foil is varnished on one side, and the action of the press cuts out and seats a full row of disks at each stroke, a trip mechanism automatically stopping the machine when the last row is finished. The primers are now ready for the insertion of the anvil, when the latter is necessary.

Inserting the Anvil. — The anvil, which is made as previously described, is now inserted into the primer cup. To do this, the anvils are shaken into plates which are of the same size as the charging plate and have the same number of holes, located, of course, in the same relative positions. The charging plate which still holds the charged and varnished cups is then covered with a plate containing the anvils, and the two plates are taken to a machine known as an "anviling" press, where the anvils are forced into and seated in the cups. The construction and operation of this anviling press is substantially the same as that used for pressing the detonating charge in the cups. After a plate full of anvils has been pressed in, the charging plate is then taken to an ejecting machine where the primers are ejected. They are then taken to a drying room to remove any moisture and thus seal the primer by means of dry varnish.

Some primer manufacturers prefer not to insert the anvil into the primers immediately after they have received the final drop of varnish, and, in this case, the charged cups are ejected from the charging plate and dried. Later they are again shaken into the plate which is assembled with the anvil plate, as previously described, and taken to the anviling press.

Wet Method of Charging Primers.— The method of charging primers with a detonating composition in the wet condition requires a primer charging table of a different type from that illustrated in Fig. 5. In this process, the primer cups are shaken into charging plates, varnished, and partially dried. Then the plate is inserted into a slot located at the rear of the charging table. The mixture is rubbed into the holes in the turn-over or charger plate, which is of the proper thickness and is provided with holes of such a diameter as to hold the correct charge of detonating composition required for each cup. The charger plate is then turned over onto the charging plate, and the pin plate located above the charger plate, which is an auxiliary member of the device, is turned down so as to force the small pellets of detonating composition out of the holes in the charger plate into the cups held in the lower plate. The charging plate with its charged cups is then removed to the "foiling" press where the disks of tin foil are punched out and pressed over the still damp priming mixture. After this operation, the anvils can be assembled in the cups or the cups may be first dried for a considerable time and the anvils assembled later, in the manner previously described.

The detonating composition used in primers for military rifle cartridges is generally of the potassium-chlorate mix·

ture, which is given on page 21. This composition is safe to handle in a wet condition, is sufficiently sensitive, when dry, and emits a large body of flame, when discharged. The large body of flame makes this composition superior to fulminate of mercury for use with smokeless powders.

INDEX

	PAGE
Annealing cartridge blanks	22
Annealing mouth of center-fire cartridges	47
Anvils, blanking and forming primer	156
inserting the primer	160
Assembling brass caps and paper shells	129
Assembling spring in cartridge clips	113
Austrian army cartridge	7
Automatic feed mechanism for brass cups	134
Battery pocket of shot shells	131, 135
Black gun powder	10
manufacture of	12
Blanking primer anvils	156
Blanking primer cups	155
Blanks for French cartridge case	143
Brass caps and paper shells, assembling	129
Brass heads for shot shells	127
British army cartridge	7
Buck-shot charges for shot shells	141
Bullet, forming and trimming French	150
making the French	149
Bullets, canneluring spitzer	101
dies for spitzer	102
drawing metal case for	55
early development	3
for center-fire cartridges	54
for rim-fire cartridges	32
making spitzer	88
sizing spitzer	101
swaging	34
swaging lead slugs for	93
Canneluring rim-fire cartridges	37
Canneluring spitzer bullets	101
Caps, manufacture of percussion	154
Cartridge cases, early development	2
Frankford arsenal method of drawing	61
Cartridge clips, and chargers	8
making	109

PAGE

Cartridge manufacture, dies for 75
Cartridge, manufacture of French. 143
Cartridges, chief requirements of. 41
   gaging, weighing, and inspecting.. 113
   history 1
   loading and clipping.. 105
   manufacture of center-fire....... 41
   manufacture of rim-fire.... 22
   modern 6
   principal dimensions 5
Casting bullets 32
Center-fire cartridges, manufacture of 41
Chargers, cartridge 8
Charges for shot shells 139
Charging primers 157
   wet method 161
Chilled and drop shot, manufacture of.. 140
Chilled shot 139
Chlorate of potassium 21
Clipping cartridges 105
Clips, cartridge 8
   inserting cartridges in.... 115
   making cartridge 109
Cluster double-action punches and dies 82
Crimping center-fire cartridges.... ......'...... . 59
Crimping rim-fire cartridges.. 37
Cupping cartridge blanks. 22
Cupping French cartridge case... 144
Cupping primer cups. 155
Cups, for cartridge cases, making. 61
   for spitzer bullets 88

Detonating composition, pressing. 160
Development of military rifle cartridges .. ' 1
Die blanks, making redrawing. 76
Dies and punches, cluster double-action...................... 82
Dies, for cartridge manufacture... 75
   for spitzer bullets.. 102
Dimensions of cartridges . . . 5
Drawing cartridge cases...... 61
Drawing metal case for bullet....... 55
Drawing operations, on center-fire cartridges 44
   on rim-fire cartridges.... 24
Drawing press feed chute 70
Drawing punches and dies for cartridge cases, setting... 65

PAGE

Drop and chilled shot, manufacture of... 140
Drop shot 139

Explosives used in cartridges 10

Feed chute for drawing press.. 70
Feed mechanism for small brass cups 134
Forming French bullet..... 150
Forming primer anvils. 156
Forming the head of center-fire cartridges... 46
Forming the head of rim-fire cartridge cases................. . 30
Frankford arsenal method of drawing cartridge cases. 61
French army cartridge 7
French bullet, making.................... . . 149
French rifle cartridge, manufacture of 143
Fulminate of mercury .. . .˙ 19

Gaging cartridges 113
German army cartridge 7
Greasing cartridges 39
Gun powder, black 10

Heading operation on French cartridge case.. 145
Heading shot shells....................... 130
Heads for shot shells. 127
History of military rifle cartridges.. 1

Indenting operations on French cartridge case 144
Inspecting cartridges .. .. -. 113
Inspecting French cartridge.. 152
Inspecting French cartridge case. 149
Inspecting shot shells 135
Italian army cartridge 7

Lapping redrawing dies. 79
Lead fillings for spitzer bullets. 90
Lead slugs for bullets, swaging.. 93
Loading cartridges 105
Loading center-fire cartridges 58
Loading French cartridge........ 152
Loading rim-fire cartridges 35
Loading shot shells 137

                                                                    PAGE
Mercury fulminate   19
Military rifle cartridges, history..   1
, manufacture of French. . . . .   ` 143
   modern   6
   principal dimensions   5
Mouth-annealing French cartridge case..   145

Nickel cases for spitzer bullets   89

Packing center-fire cartridges.   60
Packing rim-fire cartridges   39
Percussion caps, manufacture of   154
Piercing battery pocket of shot shells   135
Potassium chlorate   21
Powder, smokeless   14
   used in cartridges. . . .,.   10
Pressing the detonating composition for primers   160
Primer anvil, blanking and forming.   156
   inserting   160
Primer cups, blanking and cupping.   155
Primers, charging   157
   detonators for   19
   inserting in center-fire cartridges   .   52
   manufacture of   154
   wet method of charging   161
Priming French cartridge case   148
Priming rim-fire cartridges..   31
Priming shot shells.   135
Punches and dies, cluster double-action   82
Punches, making redrawing.   81

Redrawing dies, lapping.   79
Redrawing operations, on cartridge case.   67
   on French cartridge case.   144
   on nickel cases for spitzer bullets.   89
Redrawing punches, making   81
Redrawing center-fire cartridges.   48
Rifle cartridges, chief requirements of   41
   history   1
   manufacture of center-fire   41
   manufacture of French   143
   modern   6
   principal dimensions   5
Rim-fire cartridges, manufacture of.   22
Russian army cartridge . . . .,.   7

PAGE

Shells, manufacture of shot.. . . 117
Shot for shot shells, manufacture of . 140
Shot shells . 117
Sizing spitzer bullets... .. . 101
Sizing tubes for shot shells.. . 121
Slugs for bullets.. . 54
    casting . 32
Slugs for spitzer bullets.... . 90
Smokeless powder .. . 14
    classification of . . 18
    manufacture of . 16
Spitzer bullets, canneluring........ . 101
    dies for . 102
    making . 88
    sizing . 101
Spring for cartridge clip, making. . 112
Swaging bullets . 34
Swaging lead slugs for bullets. . 93

Tapering French cartridge case.. . 145
Testing center-fire cartridges.. . 59
Testing rim-fire cartridges.. . 39
Testing shot shells. . 137
Trimming cartridge cases. . 73
Trimming center-fire cartridges.. . 45, 50
Trimming French bullet. . 150
Trimming metal case for bullets.. . 55
Trimming rim-fire cartridge cases . 27
Tubes for shot shells, making.. . 117
Tumbling slugs for bullets. . 33

United States Army cartridge. . 7

Verifying center-fire cartridges . 49

Washing cups for cartridges. . 23
Waterproofing tubes for shot shells.. . 121
Weighing cartridges . 113
Wet method of charging primers . 161

Printed in the USA
CPSIA information can be obtained
at www.ICGtesting.com
LVHW050346261023
761875LV00030B/41